BASIC LANDSCAPE ECOLOGY

by

Robert N. Coulson
Professor
Knowledge Engineering Laboratory
Department of Entomology
Texas A&M University

and

Maria D. Tchakerian
Associate Research Scientist
Knowledge Engineering Laboratory
Department of Entomology
Texas A&M University

Copyright © 2010 Knowledge Engineering Laboratory Partners, Inc.

ISBN 978-0-9831617-0-7

Library of Congress Control Number: 2010917617

Mailing Address:

KEL Partners, Inc.
3010 Cochise Ct.
Suite 23
College Station, TX 77845
http://www.kelabpartners.com/

Cover photograph, *The Environmental Cube*, courtesy of Syngenta®.

CONTENTS

PREFACE	vii
Acknowledgments	x
Preface Part I — Foundation	xi
1. Introduction to Basic Landscape Ecology	1
What is Landscape Ecology?	1
Where Did Landscape Ecology Come From?	6
Why is Landscape Ecology Important?	9
2. Organization of the Knowledge Base for Ecology	11
Levels of Ecological Integration	11
Heirarchy Organization of Systems	16
Emergent Properties of Systems	16
Ecology in Landscapes	19
3. Spatial and Temporal Scale in Landscape Ecology	35
Spatial and Temporal Extent	36
Cartographic (Map) Scale	40
Scale in Quantitative Ecology	43
4. The Ecosystem Concept in Landscape Ecology	47
Ecosystem Structure	48
Ecosystem Function	49
Ecosystem Change	53
Ecosystem Services	59
Ecosystem Management	61

5. The Landscape in Ecology — 65
- Landscape Structure — 66
- Landscape Function — 71
- Landscape Change — 73
- Landscape Management — 79

6. Landscape in Art, Geography, and Landscape Architecture — 81
- Landscape and Art — 82
- Landscape and Geography — 85
- Landscape and Landscape Architecture — 86

Preface Part II — The Substance of Landscape Ecology — 91

7. Landscape Structure: Environment, Geometry, and Perception — 95
- The Landscape Environment — 97
- The Geometry of Landscapes — 104
- Perception of the Landscape Environment — 149

8. Landscape Function — 159
- General Model of Landscape Function — 160
- Attributes of Causal Agents that Initiate the Propagation of Ecological Effects — 166
- Entities Propagated Across the Landscape Environment — 167
- Vectors Responsible for Transmission of Entities in the Landscape Environment — 172
- Consequences of Propagation of Ecological Effects in the Landscape Environment — 177

9. Landscape Change — 185
- Landscape-Cover Change — 187
- Landscape-Use Change — 216
- Effects of Landscape Change on Living Organisms — 221
- Development of Pattern in Mosaic Landscapes — 230

10. Landscape Analysis and Synthesis — 239

    Analysis of Landscapes — 240

    Systhesis of Spatial Data and Information in Landscapes — 248

    Knowledge Engineering — 254

Glossary — 261

References — 273

Index — 295

Basic Landscape Ecology

PREFACE TO BASIC LANDSCAPE ECOLOGY

The landscape ecology enterprise is broad-based. A typical International Association for Landscape Ecology (IALE) conference is populated by scientists, tool developers, land-use managers, planners, designers, conservationists, etc. The community consists of ecologists, geographers, landscape architects, foresters, agronomists, and engineers. Interspersed within this mixture are students. Because landscape ecology is an important topic to such a large and diverse audience, the literature on the subject is immense. In the mid-1980s, when interest in landscape ecology began to flourish, the curious could begin their journey with Naveh and Leiberman (1984), Forman and Godron (1986), Zonneveld and Forman (1990), Forman (1995), Farina (2000a and b), and Turner et al. (2001). In the interim since these titles were issued, the knowledge base in landscape ecology has greatly expanded. Today, the substance of the literature includes scientific discovery in landscape ecology; methodologies for landscape description, analysis, and synthesis; and the application of the science for landscape-use management, planning, and design. This literature is dispersed among many texts and a variety of specialty journals. So where to begin a study of landscape ecology is not as clear as in the past. This text is intended to provide a starting point for the study of landscape ecology. The goal is to provide a contemporary synthesis of basic landscape ecological concepts with an applied interpretation. The expected learning outcome is (1) a broad-based knowledge of the basic concepts of landscapes ecology, (2) an understanding of the relationship between concepts of landscape ecology and landscape-use management, planning, and design, and (3) an overview of tools and techniques for spatial description, analysis, and synthesis.

This text is divided into two sections (Figure I). Section I, which consists of six chapters, is intended to provide a uniform background for students from various academic disciplines. The subjects treated in this section include: Introduction to Basic Landscape Ecology (Chapter 1), Organization of the Knowledge Base for Ecology (Chapter 2), Spatial and Temporal Scale (Chapter 3), The Ecosystem Concept in Landscape Ecology (Chapter 4), the Landscape in Ecology (Chapter 5), and Landscape Ecology in Art, Geography, and Landscape Architecture (Chapter 6). Section II, which consists of four chapters, is intended to provide an examination of the substance of contemporary landscape ecology. The subjects treated in this section include: Landscape Structure (Chapter 7), Landscape Function (Chapter 8), Landscape Change (Chapter 9), and Landscape Analysis and Synthesis (Chapter 10).

This organizational format is based on our experiences in teaching an introductory graduate course in landscape ecology at Texas A&M University. The course, which was first offered in 1988, is typically populated with a blend of students from various academic programs offered by the Colleges of Agriculture and Life Sciences, Science, Architecture, Geosciences, Liberal Arts, Veterinary Medicine, and Engineering. It is also a required course in the University-wide GIS Certificate Program. Many of the students do not have a background that includes a foundation in basic ecology and most have targeted reasons for taking the course, e.g., the ecologists may be interested in some aspect of species movement, the landscape architects in ecological planning and design, the geographers in biogeography, the veterinary students in disease epidemiology, etc. As many of the students are "visual learners," this modality is emphasized. All the students are beginning their study of landscape ecology.

Robert N. Coulson
Maria D. Tchakerian
College Station, TX
December 2010

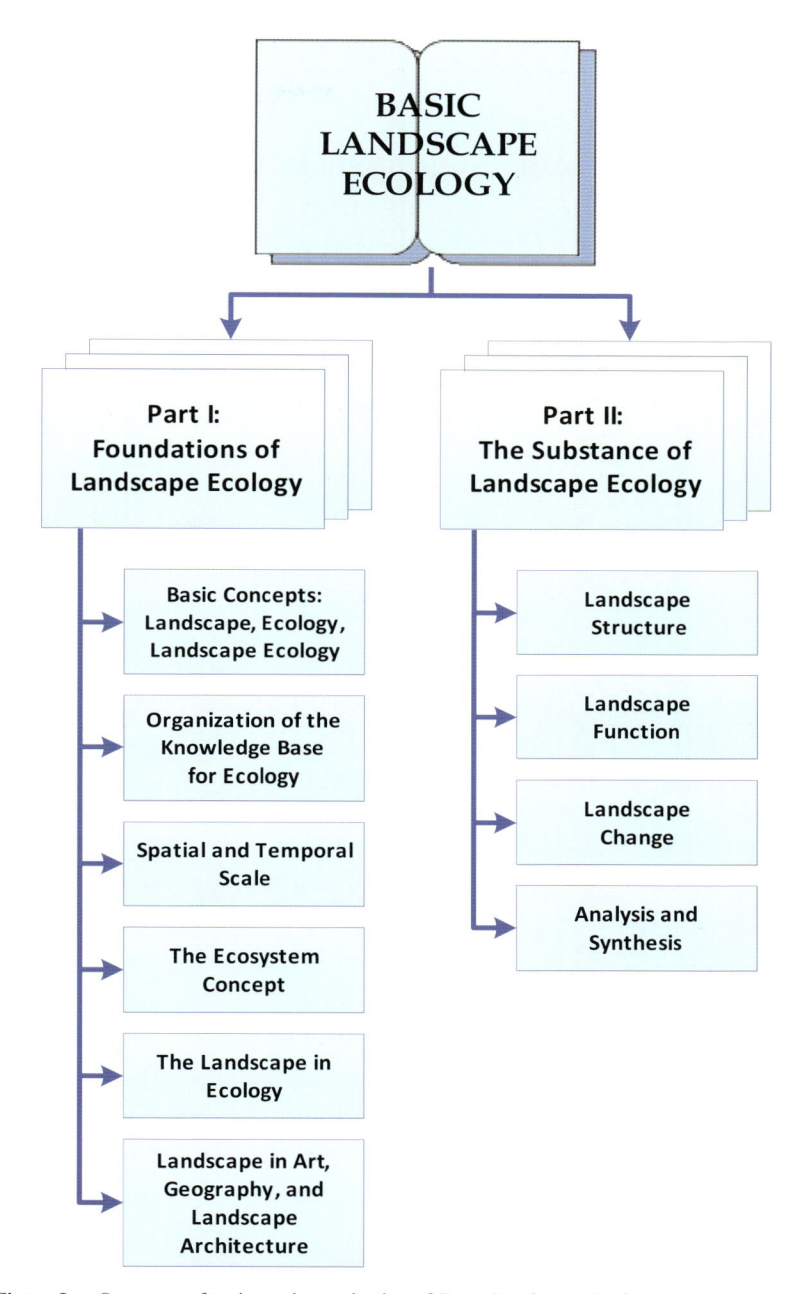

Figure I — Summary of topics and organization of *Basic Landscape Ecology*.

Acknowledgments

Many individuals contributed to the odyssey that led to the preparation, production, and publication of this text. Our partners in the Knowledge Engineering Laboratory (KEL) at Texas A&M University contributed their technical skills, shared their thoughts and ideas on the content, reviewed and critiqued the manuscript of the text, provided figures and images, and participated as subjects in our teaching experiments. In particular, we acknowledge and thank A. Birt, K. Baum, S. Bird, A. Bunting, D. Cairns, S-H. Chen, G. Curry, D. Dighe, T. Dudek, S. Fargo, R. Feldman, J. Fitzgerald, R. Flamm, R. Giardino, S. Gupta, S. Kim, C. Lafon, D. Loh, C. Lovelady, R. Meegan, L. Mussachio, S. O'Keefe, A. Pinto, S. Petty, M. Saunders, H. Saarenmaa, T. Schowalter, E. Takow, J. Waldron, D. Wunneburger, W. Xi, J. Yu, and Y. Zeng.

Special thanks are due to João Azevedo, D. Kulhavy ("Dr. Bug"), N. McIntire, and J. Waldron for their critique and editorial commentary on the manuscript of this text. The remaining errors belong to us.

We are indebted to and dedicate this book to our families (Frances Coulson, Karah and Mark Dalton, and Amy and Josh Krupa) and (Vatche and Sebastian Tchakerian).

R.N.C.

M.D.T.

Preface
Part I — Foundation

We begin the study of landscape ecology by examining basic concepts that form the foundation of the discipline. The goal of this discussion is to provide a uniform background for students from the various academic disciplines that typically populate an introductory course in landscape ecology. This section consists of six chapters (Figure I.1). In Chapter 1 we pose and answer three fundamental questions: What is landscape ecology, where did landscape ecology come from, and why is landscape ecology important? In Chapter 2, we place landscape ecology into the broader discipline of ecological science by examining three basic concepts dealing with: levels of ecological integration, hierarchy in systems, and emergent properties of systems. In Chapter 3, we address the subject of spatial and temporal scale. The study of landscape ecology requires a clear concept of scale, and we consider three perspectives: spatial and temporal extent, cartographic (map) scale, and scale in quantitative ecology. In Chapter 4, we examine the ecosystem concept. In this text we represent ecosystems as basic building blocks of landscapes. For this reason, we amplify the definition of the concept by examining ecosystem structure, function, change, services, and management. We also identify alternative uses and meanings of the concept. In Chapter 5, we provide a general overview of the landscape in an ecological context. We address landscape structure, function, change, and management. These subjects are treated in detail in Part II. In Chapter 6, we conclude with a discussion of how art, geography, and landscape architecture treat the landscape concept.

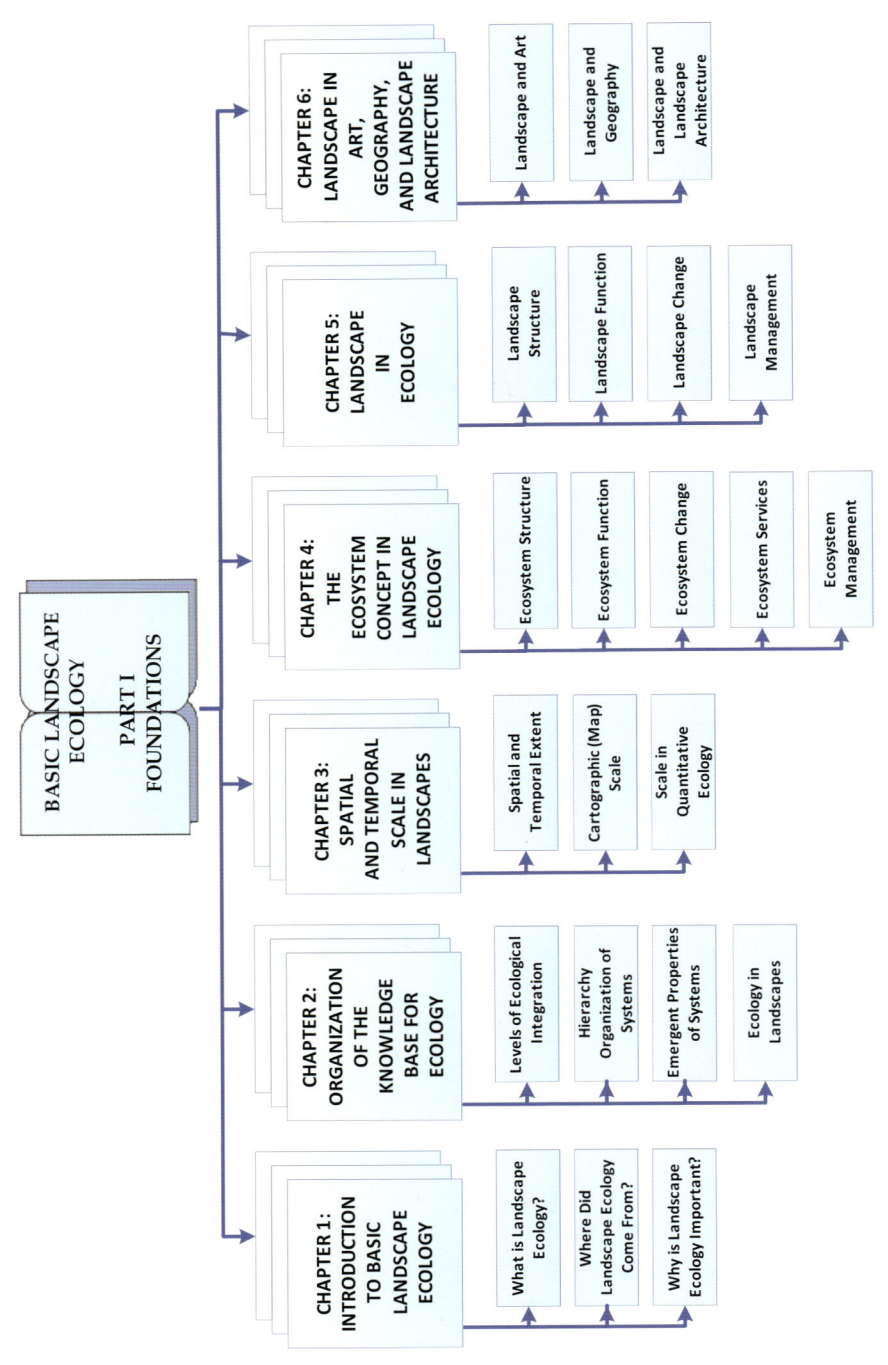

Figure I.1 — Summary of topics and organization of Part I- Foundations.

xii | *Preface to Part I*

1

Introduction to Basic Landscape Ecology

OVERVIEW

The purpose of this chapter is to launch our investigation of basic landscape ecology. We pose and answer three fundamental questions. The first question is simply: What is landscape ecology? In answering this question we introduce working definitions of *landscape, ecology,* and *landscape ecology*. The second question is: Where did landscape ecology come from? In answering this question we examine the genesis of landscape ecology, which includes consideration of both the ecological and geographical underpinning. The facilitating role of the "digital age" is also included in this discussion. The third question is: Why is landscape ecology important? In answering this question we examine how landscape ecology has broadened the scope and bounds of ecological science and also become an important enterprise that addresses human use of the landscape environment (Figure 1.1).

WHAT IS LANDSCAPE ECOLOGY?

The apparently simple question - What is landscape ecology? - leads to a variety of correct answers. The United States Regional Association of the International Association for Landscape Ecology (US-IALE) website <http://www.cof.orst.edu/org/usiale/> provides an examination

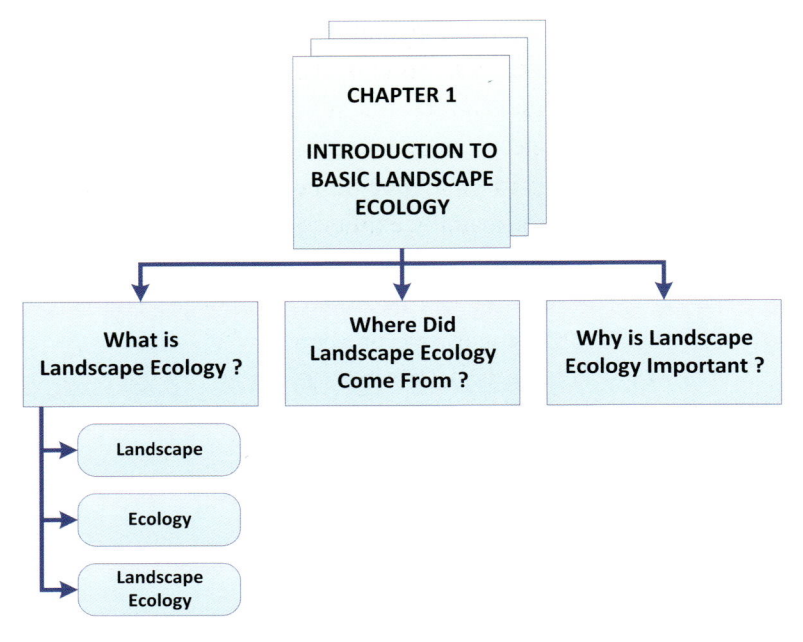

Figure 1.1 — Summary of topics and organization of Chapter 1.

of this question as viewed by ten prominent landscape ecologists. These experts share a common interest in landscape ecology, but their domain knowledge and background are somewhat different. Their views represent perspectives from ecological science, forestry, wildlife management, conservation ecology, landscape architecture, etc. This diversity results in unique and tailored definitions of landscape ecology. So our first task in addressing the subject of landscape ecology is to provide a general definition that is sufficiently clear to orient a diverse student clientele. We consider definitions of *landscape, ecology*, and *landscape ecology*. The ambiguity in the definition of landscape ecology centers on what is meant by *landscape* and Chapters 5 and 6 in Part I of this text address the subject in further detail.

Landscape

Diverse disciplines (art, geography, landscape architecture, forestry, agronomy, ecology, etc.) have claimed the term *landscape,* and consequently a variety of specialized definitions and concepts have arisen.

For our purposes, a *landscape* is a spatially explicit geographic area, i.e., an area defined by coordinates, consisting of recognizable and characteristic component entities. These entities are referred to in the literature as ecosystems, ecotopes, sites, elements, tessera, units, etc. (Figure 1.2). In tailoring the definition to your specific needs, substitute one of these terms, e.g., a landscape is a spatially explicit geographic area consisting of recognizable and characteristic component *ecosystems*. If you needed to be more precise, your definition could be restated: A landscape is a spatially explicit geographic area consisting of recognizable and characteristic component *ecotopes* (an ecotope is a bounded ecosystem) (Figure 1.2). Herein, we will generally use the term ecosystem to refer to the entities of a landscape, i.e., ecosystems are the basic building blocks of landscapes. The ecosystem concept will be examined in Chapter 4.

There are several noteworthy attributes associated with the ecological use of the term *landscape*:

1. Landscapes are typically characterized by heterogeneity - composition of parts of different kinds, i.e., they consist of more than one kind of ecosystem (ecotope, tessera, element, or unit). How many different kinds of discrete entities (ecosystems) can you identify in Figure 1.2?

2. The ecosystems that characterize a specific landscape are defined fundamentally by climatic and edaphic variables superimposed on an underlying geomorphology.

3. Characteristic patterns of landscapes result from the operation of ecological processes, i.e., "Ecological processes generate patterns, and by studying these patterns we can make useful inferences about the underlying processes" (the pattern/process paradigm, quoted from Urban 1993). In Chapter 4, we will examine the basic processess associated with ecosystem function: primary production, consumption, decomposition, and abiotic storage.

4. The ecosystems that form a landscape are arrayed as a mosaic. The spatial arrangement of the component ecosystems influences the flux of energy, materials, and information within the landscape. Both the content (kinds) and context (juxtaposition) of the ecosystems influence landscape functioning.

5. The scale (spatial extent, size) of a landscape is variable (small to large). The scale of interest can be defined in two different ways: (1) by the geometry of the component ecosystems, which is an observer-based approach, or (2) by how a specific organism (human or otherwise) responds to or perceives the environment, which is a participant-based approach.

6. Landscapes function as a consequence of the exchange of energy, materials, and information within and among the constituent ecosystems that form the mosaic of the landscape environment.

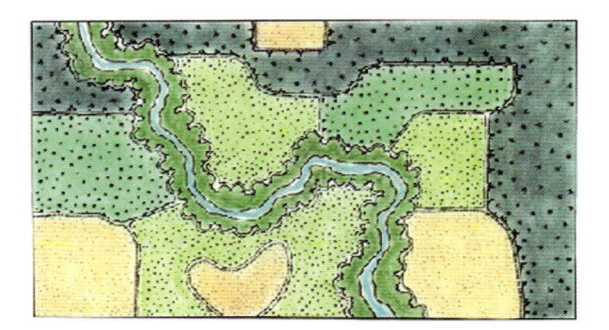

Figure 1.2 — Overhead and plan views of a forest landscape. A landscape is a spatially explicit geographic area (an area defined by coordinates) consisting of recognizable and characteristic component entities. The entities are variously referred to as ecosystems, ecotopes, tessera, elements, and units (KEL drawing).

7. Landscapes change over time as a consequence of landscape-cover change, which deals with the alteration of the biophysical attributes of the landscape environment; and landscape-use change, which deals with human purpose or intent as applied to the landscape environment. These mechanisms involve bio-ecological and geo-ecological processes (Moss 2005).

Ecology

The science of *ecology* also has a variety of definitions, but most are captured in the following statement: the study of how organisms interact with their environment (Campbell and Reece 2005). Notwithstanding the reclassification of living organisms into three Domains and six Kingdoms (Figure 1.3), this definition is remarkably similar to that provided originally by Haeckel (1869) [translated in Allee et al. (1949)]: the study

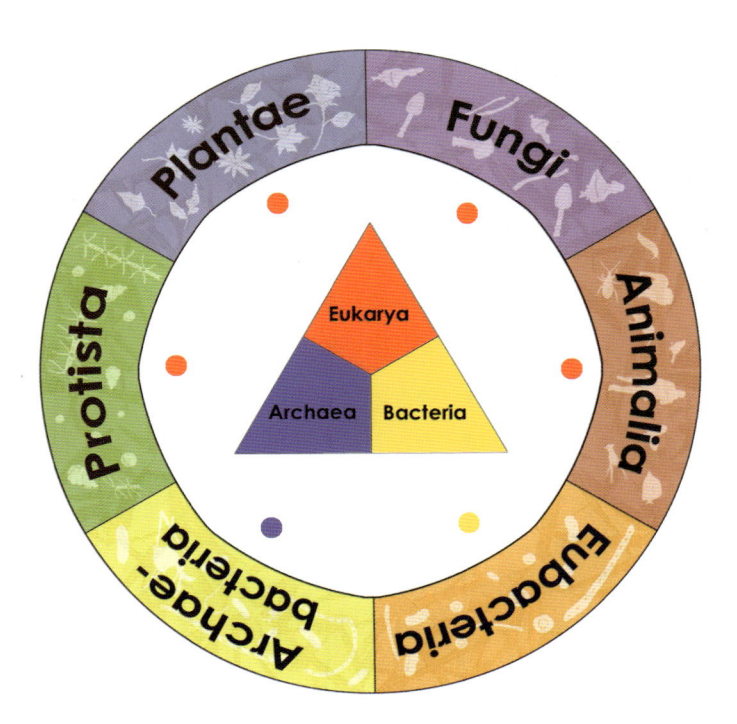

Figure 1.3 — The three Domains and six Kingdoms used in modern biology to classify living organisms. When Haeckel (1869) provided the first broadly recognized definition of ecology, the classification of life consisted of plants and animals (KEL drawing).

of the interrelations of plants and animals (i.e., living organisms) with their environment. The agenda of ecology is broad-based but much of the research emphasis centers on investigations that seek to explain the behavior and adaptations of individual organisms to their environment (the province of autecology), causes for change in the distribution and abundance of populations of organisms (the province of population ecology), the accommodations that organisms make for one another (the province of community ecology), the interrelations of the biotic [living] community with the abiotic [non-living] environment (the province of ecosystem ecology).

Landscape Ecology

Landscape ecology is the subdiscipline of the parent discipline of ecology that considers the agenda of ecology (behavior and adaptations, distribution and abundance, accommodations, and biotic/abiotic interactions) in a spatial context. So our working definition of landscape ecology is as follows: *Landscape ecology* is the science that embraces the agenda of ecology, however broadly or narrowly defined, in a spatially explicit manner. "An ecological question is a landscape ecology question if it cannot be answered without considering spatial components" (Gustafson 2006), i.e., an ecological question is a landscape ecology question if the answer requires consideration of spatial components. Our definition of landscape ecology is purposefully simple but understandable without a developed background in either basic ecology or landscape ecology. As we progress through this text, the definition will be expanded and clarified.

WHERE DID LANDSCAPE ECOLOGY COME FROM?

In comparison with most other scientific disciplines, landscape ecology has a brief history (Moss 2005, Turner 2005a and b). The plant geographer Carl Troll (1939) is credited with coining the term *landscape ecology*. From this point, the science and practice of landscape ecology grew from dialog among geographers and ecologists. Two lines of development ensued, which Moss (2005) refers to as the *geo-ecology* tradition and the *bio-ecology* tradition. The geo-ecology position developed primarily in Europe and was a product of both physical and cultural geographers. The physical geography component integrated climatology, geomorphology, pedology (soil science), and biogeography; the cultural geography counterpart dealt mostly with how humans organize, use, and regulate space (socio-ecology). The bio-ecology tradition developed primarily in the United States and was the product of biological scientists, principally ecologists. The major recent advances in landscape ecology have occurred in the bio-ecology realm (Moss 2005), but the players contributing to the development of this tradition expanded to include representatives from other disciplines, e.g., landscape architecture, forestry, agronomy, applied mathematics, etc.

When the inventions of the digital age began to be incorporated into the mainstream of scientific discovery, the discipline of landscape ecology blossomed. Landscape ecology is, in part, a computational science in that many of the fundamental research and applications questions can be addressed only through the use of computer-based technologies. Consequently, the rapid evolution of digital technologies, both hardware and software, enabled the development of landscape ecology and facilitated the use of the science in management and planning activities. In parallel with the development of the landscape ecology "tool box" (GIS; map and image processing software; remote sensing technologies; spatial analysis, synthesis, modeling methodologies; GPS, etc.), the literature of landscape ecology expanded and the pioneers, proselytes, and practitioners of the discipline began to assemble, e.g., Naveh and Leiberman (1984), Forman and Godron (1986), Zonneveld and Forman (1990), Lucas (1991), Kolasa et al. (1991), Hansen et al. (1992), Boyce (1995), Forman (1995), Hansson et al. (1995), Zonneveld (1995), Nassauer (1997), Klopatek and Gardner (1999), Mladenoff and Baker (1999), Farina (2000a and b), Sanderson and Harris (2000), Turner et al. (2001), Gergel and Turner (2002), Gunderson and Holling (2002), Gutzwiller (2002), Liu and Taylor (2002), Bissonette and Storch (2003), Forman et al. (2003), Mansourian et al. (2005), Vanslembrouck and Van Huylenbroeck (2005), Wiens and Moss (2005), Wu and Hobbs (2007), etc. Figure 1.4 provides a flavor of the technological and intellectual events that contributed to the genesis of landscape ecology.

BENCHMARK EVENTS FOR LANDSCAPE ECOLOGY

Ecology

E.P. Odum, 1953.
Fundamentals of Ecology,
Published

1950

1960

Rachel Carson, 1962.
Silent Spring,
Published

1970

Ecosystem Ecology

Bormann and Likens, 1979.
Pattern and Process in a Forested Ecosystem, Published

Landscape Ecology

1980

Computer Science and Related Digital Technologies

The Founding of IALE, 1981.
Plestany, Czechoslovakia (Slovakia)

Allerton Park Conference, 1983.

IBM PC, 1981.

Naveh and Lieberman, 1984.
Landscape Ecology,
Published

Apple Macintosh, 1984.

Forman and Godron, 1986.
Landscape Ecology,
Published

Geographic Information Systems and Map/Image Processing Software

The Founding of US-IALE, 1986.
Athens, Georgia, USA

P.A. Burrough, 1986.
Principles of GIS Systems for Land Resources Assessment,
Published

Landscape Ecology, 1987.

Computer Science and Related Digital Technologies (cont.)

1990

ESRI: ArcView 1.0 released, 1992.

Internet Society Founded, 1991.

Spatial Analysis, Synthesis, and Modeling

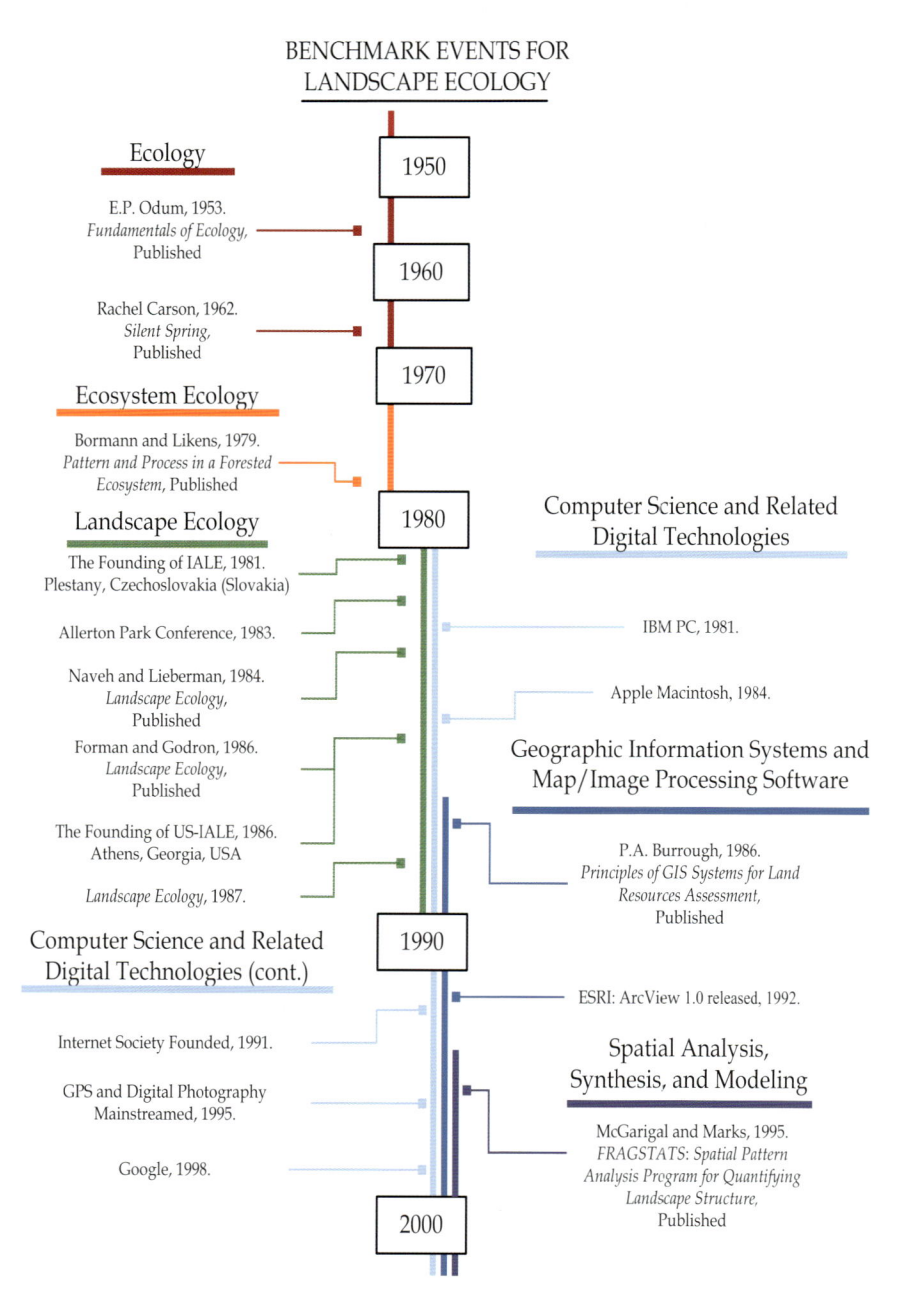

GPS and Digital Photography
Mainstreamed, 1995.

McGarigal and Marks, 1995.
FRAGSTATS: Spatial Pattern Analysis Program for Quantifying Landscape Structure,
Published

Google, 1998.

2000

Figure 1.4 — Flavors of the technological and intellectual events that contributed to the genesis of landscape ecology. The enabling technologies are products of the digital age. These products served as a stimulus for the intellectual development of landscape ecology (KEL image).

8 | *Introduction to Basic Landscape Ecology*

WHY IS LANDSCAPE ECOLOGY IMPORTANT?

Ecology became a common element in the curriculum of the biological sciences largely as a result of the seminal textbook, *Fundamentals of Ecology*, by E. P. Odum (first published in 1953 with subsequent editions). This book also broadcast the concept of ecosystem and emphasized the interconnectedness of biotic and abiotic components of the environment. Soon thereafter, Rachael Carson's *Silent Spring* (1962) (a book drawing public awareness to the dangers associated with pesticide use) triggered the era of "environmentalism" which had an agenda with a scientific, social, and political underpinning. Many looked to the science of ecology to provide solutions to problems associated with environmental management (Lubchenco et al. 1991, Coulson and Stephen 2006). But, most research within the various branches of ecology focused primarily on scientific discovery. Furthermore, few of the individuals involved in ecological research had interest or training in the application of knowledge for management purposes.

Landscape ecology has broadened the scope and bounds of ecological science and become an important enterprise for several reasons. First, because of the *geo-ecology* and *bio-ecology* traditions, the agenda of landscape ecology, *de facto*, includes human presence, actions, and activities as fundamental considerations. Second, in addition to research (the discovery of new knowledge), landscape ecology also includes activities associated with development (the integration and interpretation of new and existing knowledge) and application (the directed use of knowledge for planning, problem-solving, and decisionsupport). Third, because of this broad-based research, development, and application (RD&A) agenda, landscape ecology provides a forum for interdisciplinary study and exchange. Many of the issues addressed in landscape ecology require expertise from a variety of disciplines and consequently scientists and practitioners from diverse subject domains are brought together. Landscape ecology provides the arena for ecologists, landscape architects, geographers, civil engineers, foresters, agronomists, applied mathematicians, etc. to practice their trades together and thereby address problems that go beyond the technical expertise and domain knowledge of the individual players. Finally, landscapes represent the geographic unit that is amenable to management. Forestry, agriculture, range management, urban/suburban development, park and recreational area design and maintenance, etc., all involve modification and manipulation of the component ecosystems that make up these different landscapes. The concepts, and perhaps principles, that follow from RD&A in landscape ecology will serve to guide landscape-use management, planning, and design; and also provide a science-based foundation for these activities (Dale et al. 2000, Karr 2002).

EPILOGUE TO THE INTRODUCTION OF LANDSCAPE ECOLOGY

Our first task in introducing landscape ecology was to provide working definitions of *landscape* and *landscape ecology*. Accordingly, we defined a *landscape* to be a spatially explicit geographic area, i.e., an area defined by coordinates, consisting of recognizable and characteristic component entities. *Landscape ecology* is the sub-discipline of the parent science of ecology that considers the agenda of ecology (i.e., the behavior and adaptations of individual organisms, the distribution and abundance of populations of organisms, the accommodations that communities of organisms make for one another, and the interactions of the biotic community with the abiotic environment) in a spatial context. The discipline of landscape ecology, as a scientific endeavor, is recent and evolved from dialogue among ecologists and geographers. Advances in landscape ecology were accelerated by the inventions of the digital age that facilitated use of spatial information for scientific discovery as well as applications in landscape-use management, planning, and design. Landscape ecology is an integrative science and practice that brings together domain specialists from a variety of academic interests. Landscape ecology, *de facto*, includes human presence, actions, and activities as fundamental considerations.

2

Organization of the Knowledge Base for Ecology

OVERVIEW

Before proceeding with an investigation of the specific elements of the knowledge base of landscape ecology, we examine three general concepts which are useful in framing our discussion. These concepts include: levels of ecological integration, hierarchy, and emergent properties. A convenient and common way to organize the subject content of ecology is through levels of integration. Each level has an associated set of topics that collectively represents the agenda of ecology. Landscape ecology examines this agenda in a spatially explicit manner. Examples are provided to illustrate how subjects typically associated with the population, community, and ecosystem levels of ecological integration become landscape ecology when they are viewed in a spatial context. Hierarchy concepts are useful for examining the relationships among the different levels of integration. Each of the levels of integration also has unique emergent properties (Figure 2.1).

LEVELS OF ECOLOGICAL INTEGRATION

A *level of integration* can be viewed as a compartment containing specific information, concepts, and principles that collectively represent the knowledge base for a subject of interest. The usual progression in

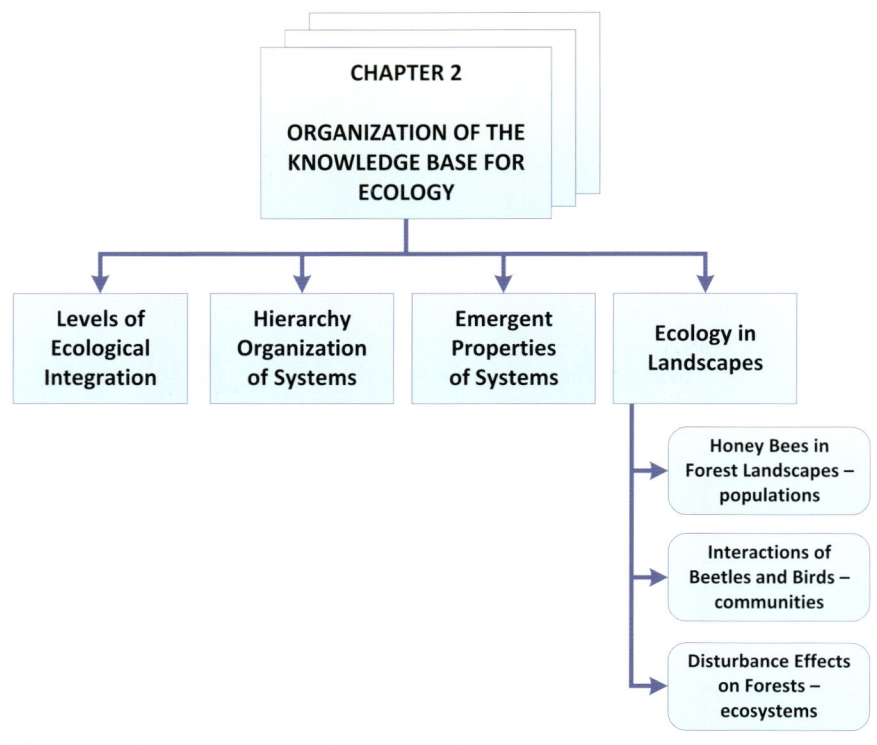

Figure 2.1 — Summary of topics and organization of Chapter 2.

the levels of ecological knowledge integration (Figure 2.2) includes the following: *individual* (a living organism), *population* (a collective group of individuals of one species that inhabits an area sufficiently small to enable interbreeding and functions as part of a biotic community), *community* (any assemblage of populations of living organisms in a prescribed area or habitat), *ecosystem* (the biotic community plus its abiotic environment), *landscape* (a spatially explicit geographic area consisting of recognizable and characteristic component ecosystems), *ecoregion* (*biome*) (a biogeographical region or formation), and the *biosphere* (that part of the earth and atmosphere capable of supporting living organisms). Numerous texts examine the substance of each of these levels of integration in varying degrees of detail. Each level of integration has a unique knowledge base associated with it. Following is an abstract of the common elements associated with the population, community, ecosystem, and landscape levels of integration.

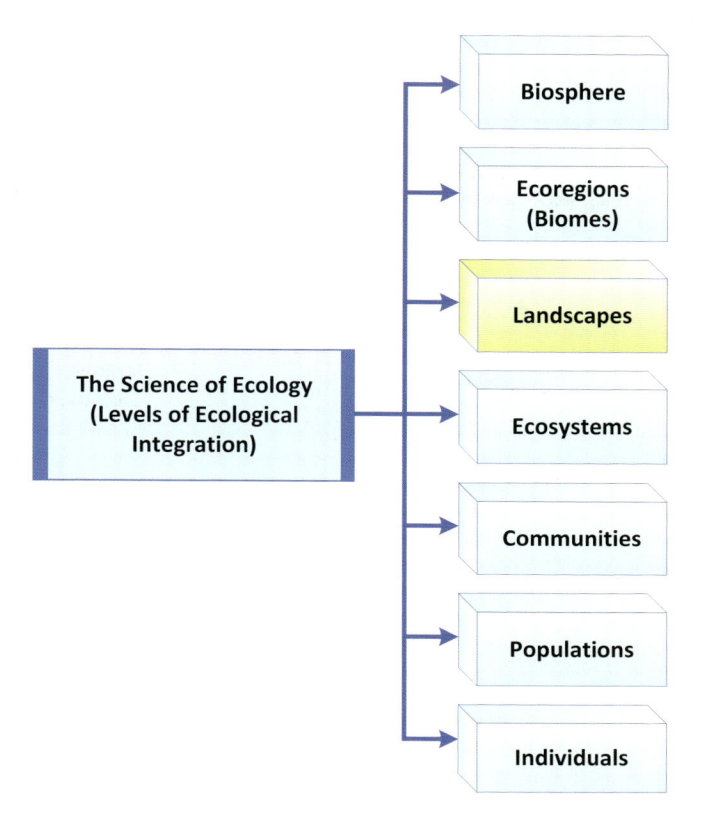

Figure 2.2 — The science of ecology organized by levels of integration. A level of integration can be viewed as a compartment containing specific information, concepts, and principles that collectively represent the knowledge base for a subject of interest. The usual progression in the levels of ecological integration includes individuals, populations, communities, ecosystems, landscapes, biomes (ecoregions), and the biosphere.

Most textbooks addressing the *population* level of integration would include a discussion of at least the following topics: properties of individuals, properties of the environment, the population processes (natality [birth *rate*], mortality [death *rate*], and dispersal), the population state variables (density, age distribution, genetic composition, etc.), adaptive evolution, and co-evolution. Imbedded in the discussion would be the causes for change in the state variables. Figure 2.3 illustrates these topics and their interrelation. The usual suite of topics for discussion in the *community* level of integration includes: niche concepts, interactions among populations and the accommodations they make for one another (herbivory, predation, parasitism, competition, symbiosis, commensalism,

Organization of the Knowledge Base for Ecology | 13

and mutualism), relative abundance, species diversity, food webs, trophic structure, guild structure, and functional groups, etc. (Figure 2.4). The discussion of the *ecosystem* level of integration includes the basic processes of primary production, consumption, decomposition, and abiotic storage; nutrient cycling; energy flow; perhaps information flow; ecological succession; ecosystem services; and ecosystem management (Figure 2.5). The *landscape* level of ecological integration typically includes a discussion of landscape structure, function, change, and management. Ecologists who specialize in a level of integration would certainly expand the agenda, as we have done for landscapes. Figure 2.6 defines in detail the specific topics we include in this level of ecological integration.

Note that in Figure 2.2 we have illustrated each level of integration as an open box or compartment. The intended metaphor is that the "stuff"

THE POPULATION SYSTEM MODEL

Co-Evolution
Environmental Modification
(Negative Feedback)

Properties of the Environment:
Resources
Conditions

Interactions

Properties of the Individuals:
Genotypes
Phenotypes

Population Processes:
Natality
Mortality
Dispersal

Population State Variables:
Density
Distribution
Age Structure

Adaptive Evolution
(Negative Feedback)

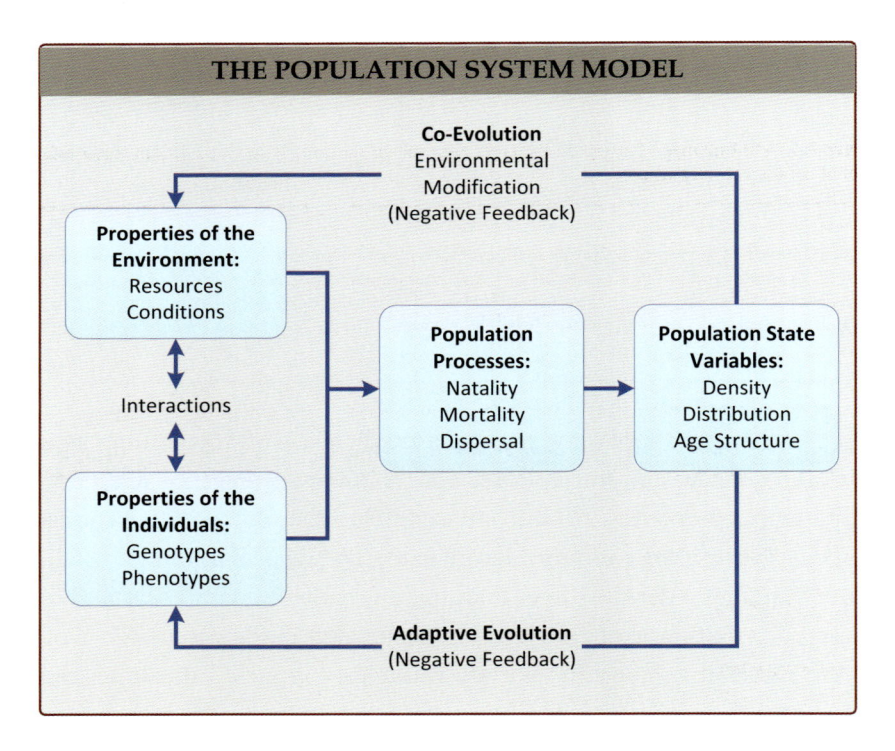

Figure 2.3 — Each level of ecological integration has a unique knowledge base associated with it. This "mind map" illustrates the basic information (elements of the knowledge base) that would be included in most discussions of the population level of ecological integration (KEL image).

14 | *Organization of the Knowledge Base for Ecology*

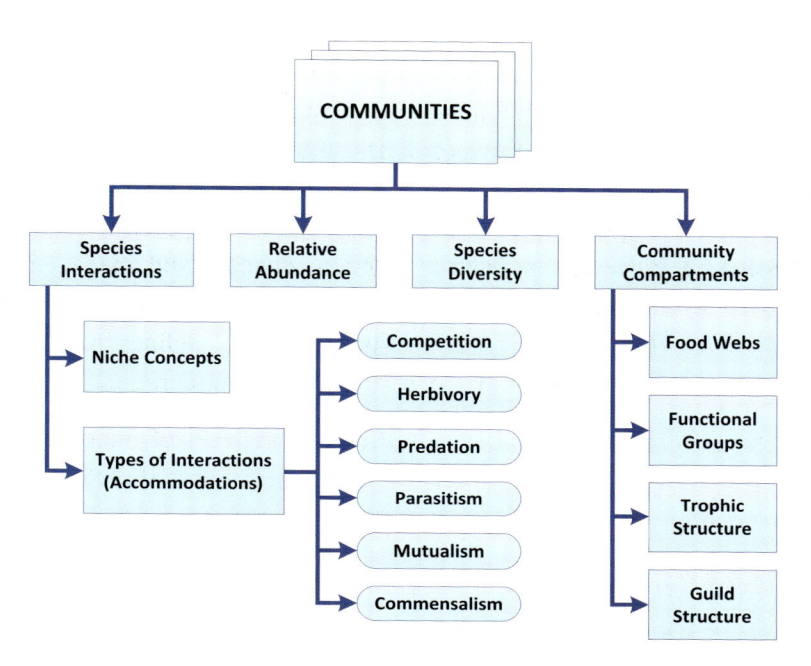

Figure 2.4 — "Mind map" illustrating the basic elements of the knowledge base for the community level of ecological integration.

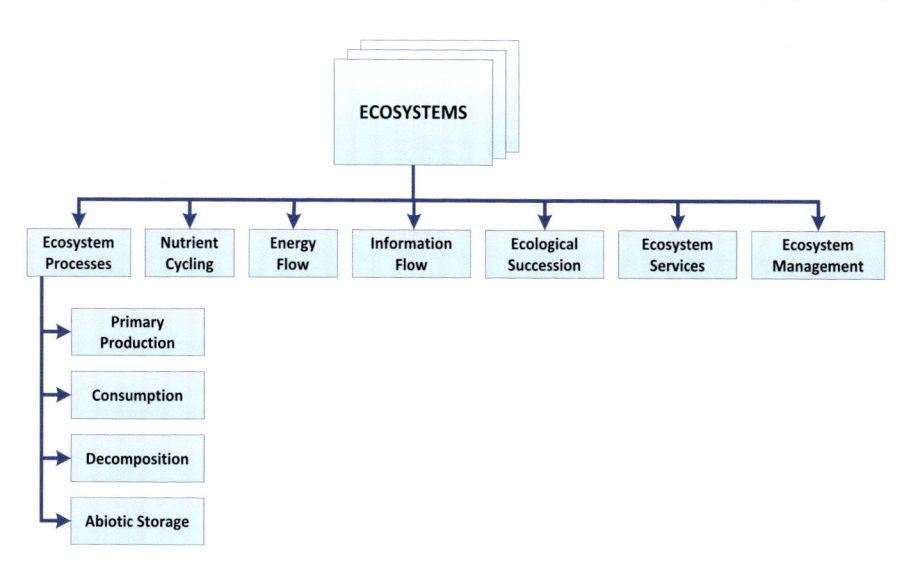

Figure 2.5 — "Mind map" illustrating the basic elements of the knowledge base associated with the ecosystem level of ecological integration.

Organization of the Knowledge Base for Ecology | 15

(subject matter) of each box can be moved around, added to, or subtracted from. But each level of integration has characteristic information, concepts, and principles associated with it. Below, we provide three examples to illustrate landscape ecological studies involving subjects associated with the population, community, and ecosystem levels of integration. What makes these studies landscape ecology is their tie to a spatially explicit geographic area consisting of recognizable and characteristic component ecosystems.

HIERARCHY ORGANIZATION OF SYSTEMS

The levels of ecological integration (Figure 2.2) are presented as a hierarchy. A *hierarchy* is a series of consecutively subordinate categories forming a system of classification. Typically, there are three levels in the hierarchy that are of particular interest (Allen and Starr 1982, Ahl and Allen 1996). The *system* itself, which is the level of study or focus; the *sub-system*, which is the level used for explanation; and the *supra-system*, which is the level used for interpretation (Figure 2.7).

In our study of landscape ecology, the system of focus or study will be the landscape. We will look to the ecosystem as the principal sub-system that will aid in explanation of structure, function, and change in landscapes. The ecoregion will be our supra-system, but herein we will not attempt much interpretation at this level. The ecoregion level of integration has been well-worked by Bailey (1996, 1998, 2002).

EMERGENT PROPERTIES OF SYSTEMS

A consequence of hierarchical organization is that as components, or subsets, are combined to produce larger functional wholes, new properties emerge that are not present at the level below. In ecology, these properties are referred to as *emergent properties*, i.e., properties of the whole not reducible to the sum of the properties of the parts. By contrast, if a property can be explained fully by mechanisms examined at the next lower level of ecological complexity, then the property is *collective* (Sanderson and Harris 2000).

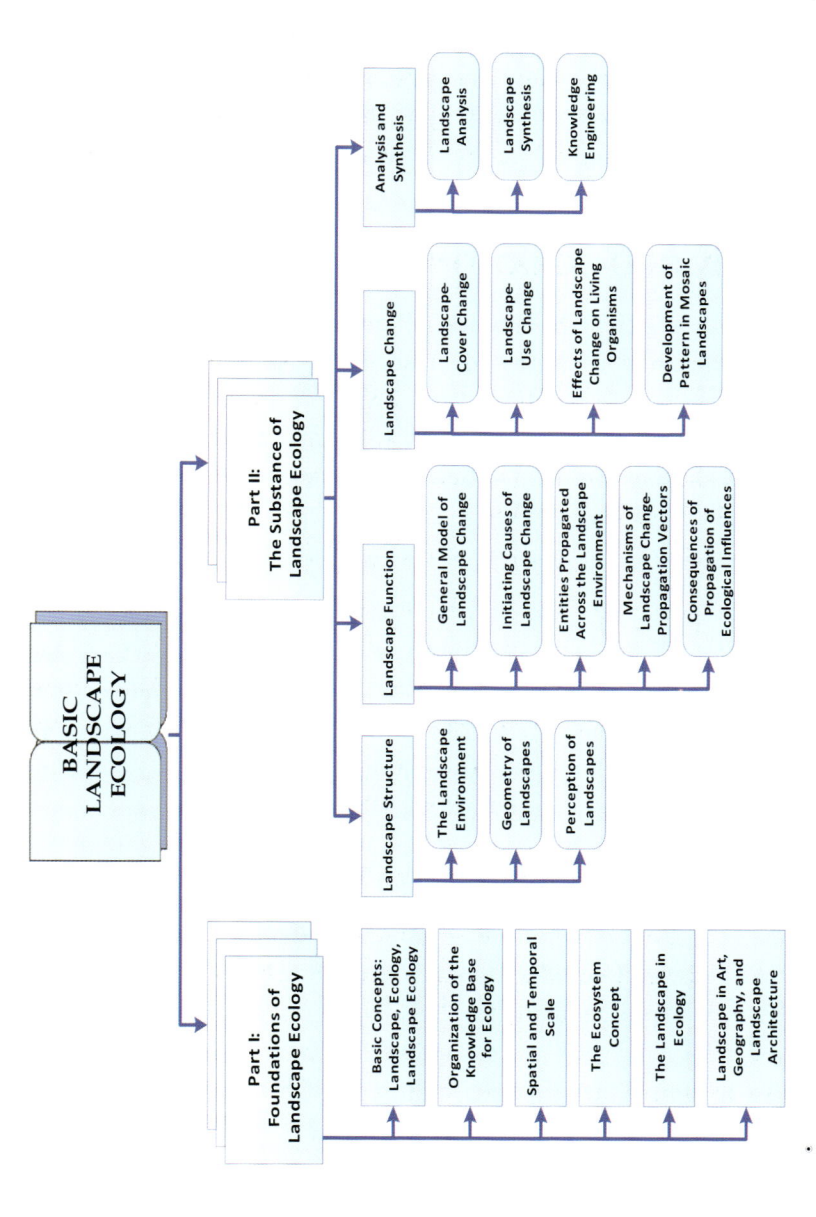

Figure 2.6 — "Mind map" illustrating our view of the basic elements of the knowledge base associated with the landscape level of ecological integration. This figure defines the topics and organization that we include as the foundation and substance of landscape ecology.

Organization of the Knowledge Base for Ecology | 17

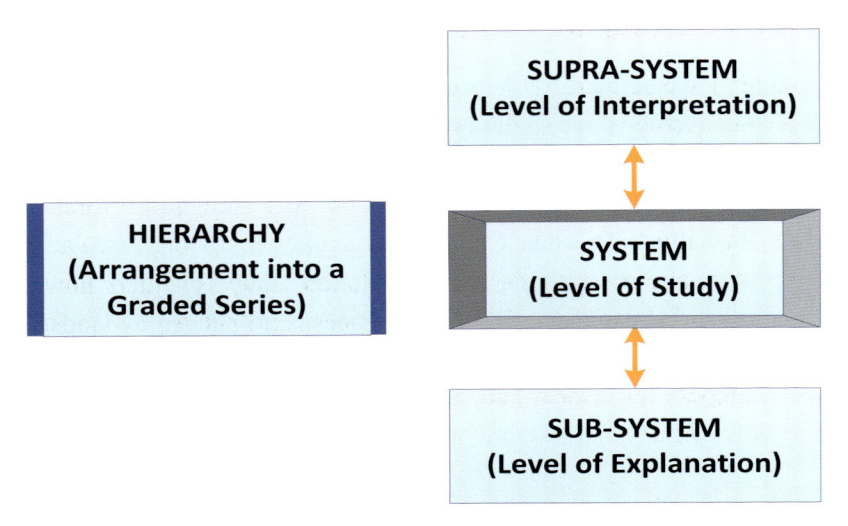

Figure 2.7 — A hierarchy illustrating three levels of structure. The level of focus is the *system* of interest. The lower level (*sub-system*) provides the explanation for how the system works, i.e., it deals with mechanism. The higher level *supra-system* is used for interpretative purposes, i.e., it deals with the significance of the operation of the system. Herein, the level of focus (the system of study) is the *landscape*, the sub-system is the *ecosystem*, and the supra-system is the *biome* (ecoregion).

Each level of ecological integration has associated emergent properties. For example, an emergent property of a population is *age structure* (*distribution*), i.e., the proportion (or percent) of individuals in a population occurring in specified age classes. The ratio of the various age groups in a population determines the current reproductive status of a population and is indicative of its future size. An emergent property of a community is *species diversity*, i.e., a property of a community that relates both the species richness (number of species present) and relative abundance (the number of individuals associated with each species). Prominent emergent properties of ecosystems include *energy flow* and *nutrient cycling* among the biotic and abiotic components. When multiple interacting ecosystems are combined to form a landscape, structural heterogeneity is introduced. *Heterogeneity* is an emergent property of a landscape. It is more than an additive property, i.e., landscapes may be identical in terms of kinds of ecosystems present but function quite differently depending on the size, shape, number, and juxtaposition of components.

ECOLOGY IN LANDSCAPES

In Chapter 1 we defined *landscape ecology* as the science that embraces the agenda of ecology in a spatially explicit manner. Above, in a simplified manner, the agenda of ecology at the population, community, ecosystem, and landscape levels of integration was introduced. Following are three case histories that illustrate landscape ecology studies involving populations, communities, and ecosystems. The population study considers how the distribution and abundance of feral honey bees is influenced by landscape structure. The community study examines how two seemingly unrelated species compete for critical habitat. The ecosystem study considers how disturbances influence nutrient dynamics within a forest landscape.

Honey Bees in Forest Landscapes – Populations

When European settlers arrived in North America, one disheartening discovery they made was that there were no honey bees, *Apis mellifera* (Figure 2.8). This discovery impacted the settlers in two important ways. First, no substantive source of sweetener was available for them. Imagine a diet without sweeteners. Second, hive products were also missing. Beeswax was of particular concern as it was commonly used for waterproofing, as a lubricant, and as a sealant. Not long thereafter, honey bees were introduced to the Americas from Europe and eventually six races became established. An additional race, the infamous African honey bee (*A. m. scutellata*) was later added to the mix of races (Pinto et al. 2004, 2005). With the development of modern agriculture, which

Figure 2.8 — The honey bee, *Apis mellifera*, was introduced into North America. This insect plays a central role in modern agriculture and is a common species in rural and urban landscapes (photograph by Scott Bauer, USDA/Agricultural Research Service).

included cultivation of large acreages with high plant density, domesticated honey bees became essential for pollination of many staple food plants. Feral honey bees, escapees from managed colonies, also became a common element of both urban and rural landscapes (Baum et al. 2008). In the 1990s scientists and practitioners noticed a decline in pollinators in general and feral honey bees in particular (Buchmann and Nabhan 1996), and this observation triggered research to investigate the problem.

In a study undertaken to examine the interaction and persistence of feral European and Africanized honey bee races, Coulson et al. (2005) monitored populations occurring in a mesoscale (100-1 000 000 ha) pine forest landscape in East Texas. Honey bees were plentiful in this landscape. This observation was surprising in that a forest landscape dominated by coniferous tree species (principally *Pinus* spp.) would not be expected to support large populations of honey bees, as food resources (both nectar and pollen) are scarce and cavity sites are rare (Figure 2.9).

The fundamental question was: How can honey bees subsist in an environment seemingly devoid of critical resources? To answer this question, the detailed structure of the landscape surrounding honey bee colonies was evaluated. Using the coordinates of established honey bee colonies as the geographic centroid, a 2 km radius was delineated, and the landscape structure within this 1256 ha unit was examined (Figure 2.10). The evaluation of landscape structure was accomplished

Figure 2.9 — A Red-cockaded woodpecker (*Picoides borealis*) cavity exploited by a colony of feral honey bees, escapees from managed colonies. Cavity sites in pine forests are rare and the Red-cockaded woodpecker is considered to be a "keystone species" in this landscape, as it is one of the few organisms that actively excavates cavities. This colony will not survive the high summer temperatures of East Texas (KEL image).

using a GIS; a classified digital spatial database consisting of multiple themes for the forest (vegetation type and cover, hydrography, ownership, wildlife management areas, wilderness areas, etc.); and a spatial analysis package, FRAGSTATS (McGarigal and Marks 1995). A variety of patch metrics were used to characterize the immediate landscape environment around honey bee colonies, e.g., number, mean size, mean shape index, density, and richness (see Table 2.1 to inspect the values for the measured variables).

Although classified as a pine forest, management practices and human activities had dissected, perforated, and fragmented the landscape. The metrics used to characterize the landscape actually defined a heterogeneous environment for honey bees that included food and habitat resources needed for survival, growth, and reproduction. Forestry practices associated specifically with road corridor maintenance, stream

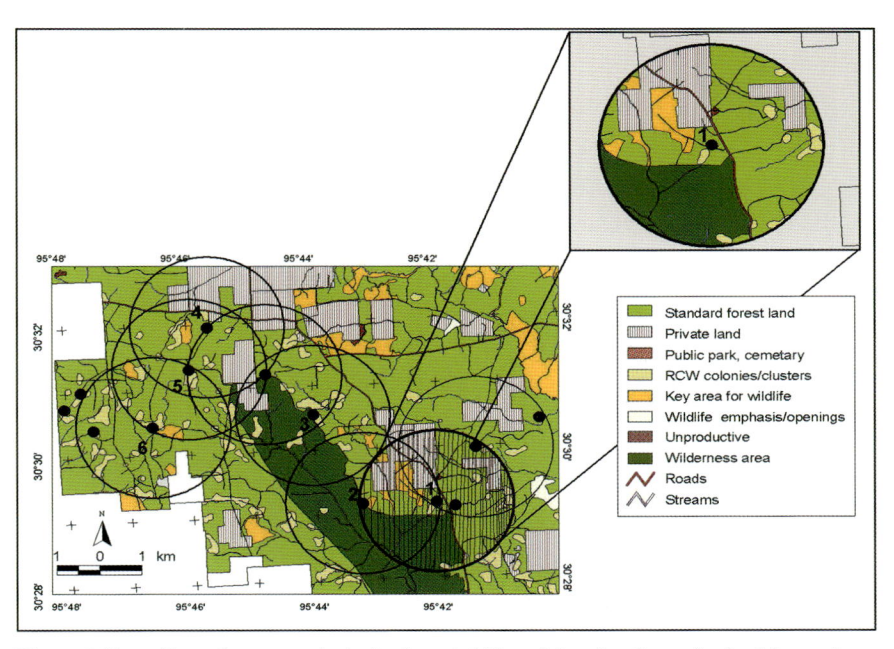

Figure 2.10 — To evaluate quantitatively the suitability of the pine forest for feral honey bees, the landscape structure surrounding established colonies was examined. The study site was on the Sam Houston National Forest in East Texas. Using the coordinates of each occupied colony as the geographic centroid, a 2 km radius was delineated and evaluated. The metrics used to characterize the 1256 ha units included the kind of patches (PR), the number (NP), the size (MPS), the shape (MSI), and the configuration (PD) of elements forming the landscape. The results of the analyses are illustrated in Table 2.1 (KEL image).

Table 2.1 — Values of metrics describing the immediate environment around honey bee colonies in the Sam Houston National Forest, East Texas. Measurements included the kinds of patches (PR), the number (NP), the size (MPS), the shape (MSI), and configuration (PD) of elements forming the landscape (McGarigal and Marks 1995).

Trap Number	Number of Patches	Mean Patch Size (ha)	Mean Shape Index	Patch Density (NP/ha)	Patch Richness
1	86	14.590	2.548	6.852	9
2	61	20.550	2.568	4.866	9
3	71	17.668	2.389	5.659	8
4	65	19.210	2.587	5.205	8
5	93	13.508	2.298	7.402	8
6	113	10.542	2.380	9.485	5

side corridor protection, wildlife habitat management, and wilderness area management introduced structural heterogeneity to the coniferous forest landscape that enriched the diversity of early successional flowering plants and provided cavity sites needed by honey bees. Ranching, farming, and urbanization within the forest also created these conditions. Furthermore, analysis of pollen extracted from local honey indicated that honey bees provided pollination services to a broad representation of native and introduced flowering species within the forest. This example illustrates an agenda item from population ecology, namely the effects of food and habitat resources on persistence of populations of honey bees, in a spatial context.

Interaction of Beetles and Birds – Communities

The southern pine beetle, *Dendroctonus frontalis*, and the Red-cockaded woodpecker, *Picoides borealis*, are indigenous species that occur together in pine (*Pinus* spp.) forests of the southern United States (Figures 2.11 and 2.12). For different reasons, these organisms have been the focus of considerable research. Interest in the southern pine beetle centers on the role this insect plays as a pest species. The insect infests southern yellow pines (*P. taeda* [loblolly pine], *P. echinata* [shortleaf pine], *P. elliotti* [slash pine], and occasionally *P. palustris* [longleaf pine]), and is the most significant mortality agent affecting softwood production in the South (Figure 2.13). It is also an important disturbance agent that serves to truncate forest development by selectively infesting senescent old-growth pines.

Figure 2.11 — Southern pine beetle adult, *Dendroctonus frontalis*. This insect, which is about the size of a grain of rice, infests and kills yellow pines in the southern United States (photograph by D. T. Almquist, University of Florida).

Figure 2.12 — Red-cockaded woodpecker, *P. borealis*. This endangered species excavates cavities in old-growth pines. The cavity trees are often colonized by the southern pine beetle (photograph by Jim E. Johnson).

Interest in the Red-cockaded woodpecker centers on the role this bird plays as an endangered species. Populations of the Red-cockaded woodpecker and other cavity-dwelling species declined dramatically as a result of logging in Southern pine forests. Consequently, much of the research on the Red-cockaded woodpecker has been directed to understanding the natural history and population ecology of the bird. A primary goal of the research was to help recover the species through habitat restoration and protection (Coulson et al. 1999).

Organization of the Knowledge Base for Ecology | 23

Rudolph and Conner (1995) reported that southern pine beetle infestations of Red-cockaded woodpecker single-cavity trees, cavity-tree cluster areas, and foraging habitat were a problem of paramount significance to the conservation and recovery of the bird. To address this problem Coulson et al. (1999) investigated the spatial interaction of populations of the two species in a forest landscape mosaic. Well-studied organisms such as the southern pine beetle and Red-cockaded woodpecker typically have extensive knowledge bases associated with them. This extant knowledge dealing with demographics and behavior was used to formulate rules describing how each organism perceives and responds to the ensemble of ecosystems forming the forest landscape environment. The specific ecosystems represent the habitat patches that contain the resources and conditions necessary for survival, growth, and reproduction. *Habitat* is the natural environment where an organism is usually found. Both the content (kinds) and context (juxtaposition) of the ecosystems forming the forest landscape were important, as human-caused fragmentation and natural disturbances created a mosaic pattern that influenced the persistence of both organisms.

Figure 2.13 — A southern pine beetle infestation on the Little Lake Creek Wilderness Area, Sam Houston National Forest in East Texas. The red and yellow trees have been colonized and killed by the insect (KEL image).

The habitat targets utilized by the southern pine beetle included acceptable yellow pine hosts, susceptible habitat patches (generally host species 40 years of age and older), and lightning-struck hosts. Lightning-struck hosts play an important role in the population dynamics of the southern pine beetle as they serve as epicenters for the initiation of infestations, refuges for dispersing beetles, and stepping stones that link populations in different habitat patches. Resin volatiles, produced as a consequence of the lightning strike, attract the beetles, which colonize and reproduce within the trees. The habitat targets for the Red-cockaded woodpecker are based on behavior associated with nesting, roosting, and foraging. Targets for nesting and roosting include the same old-growth yellow pines used by the southern pine beetle. However, when available, longleaf pine is preferred by the bird. Mature living trees are essential, as the birds excavate "resin wells" around the entrance to the cavity (Figure 2.14). The resin exuded from the tree serves as a barrier to predatory rat snakes (*Elaphe* spp.). The volatile terpenes associated with the resin may also serve as olfactory cures to dispersing southern pine beetles. Targets for foraging include living yellow pine and pine-hardwood patches 30 years of age and older within the home range of the bird (50-400 ha).

So the fundamental landscape ecology questions were: What is the nature of the spatial interaction of southern pine beetle and the Red-cockaded woodpecker populations, and what is the consequence of the interaction? To answer these questions, Coulson et al. (1999) examined the distribution and abundance of beetle and bird populations occurring in

Figure 2.14 — Artificial cavity inserted into a loblolly pine to provide a nesting site for the Red-cockaded woodpecker. Note the resin exudates around the face of the cavity. The resin exuded from the tree serves as a barrier to predatory rat snakes (*Elaphe* spp.). The volatile terpenes associated with the resin may also serve as olfactory cues to dispersing southern pine beetles (photograph by Jim E. Johnson).

a mesoscale (100-1 000 000 ha) pine forest landscape, the Homochitto National Forest in Mississippi. As with the honey bee example, the examination was accomplished using a GIS, a classified digital spatial database consisting of multiple themes for the forest (vegetation type and cover, hydrography, ownership, point locations for southern pine beetle infestations and lightning-struck host trees, coordinates of Red-cockaded woodpecker colony sites, etc.), a spatial statistical procedure (referred to as the "weighted connectivity index," (Coulson et al. 1999)), and the rules for habitat selection and use by the beetle and bird. The spatial statistical procedure measured the connectivity of the landscape according to the importance of specific landscape elements for the southern pine beetle and Red-cockaded woodpecker. Application of the procedure resulted in maps that characterized heterogeneity of the landscape as perceived by the bird and beetle, based on the rules defined for habitat selection. These maps were referred to as "functional heterogeneity" maps. An individual map was produced for the southern pine beetle (Figure 2.15) and the Red-cockaded woodpecker (Figure 2.16). By overlaying the maps, the area of interaction of the beetle and bird was identified (Figure 2.17) (Coulson et al. 1999).

The observation made by Rudolph and Conner (1995), relative to the impact of the southern pine beetle on Red-cockaded woodpecker cavity trees, cluster areas, and foraging habitat, can be explained by the fact that the two organisms perceive and respond to the same structural elements of the forest landscape mosaic; albeit for different reasons, that is to say, there is a spatial and temporal coincidence of the insect and the bird within the landscape (Coulson et al. 1999). An interesting anecdote to this investigation is that the degree of interaction between the beetle and bird can be radically altered by changing the composition of tree species in the forest landscape. In particular, if longleaf pine is a predominant tree species in the landscape, the interaction between the bird and beetle is minimized. Longleaf pine produces a significantly greater resin yield than the other species of yellow pines. Resin production by hosts is considered to be a primary defense mechanism of pines against colonization by bark beetles, i.e., the resin flow prevents colonization by "pitching" the beetles out. Southern pine beetle infestations occur much less frequently in longleaf pine forests. Given a preference, the Red-cockaded woodpecker

Bark Beetle - Functional Heterogeneity

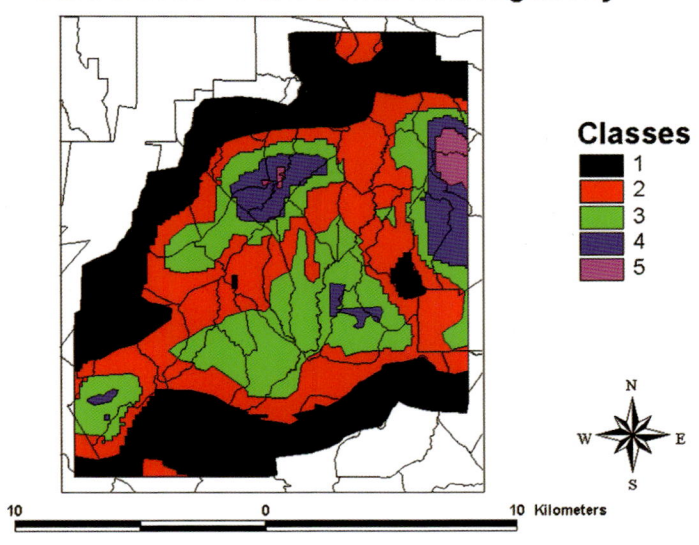

Figure 2.15 — "Functional heterogeneity" map illustrating habitat preference for the southern pine beetle (1=least suitable habitat, 5=highly suitable habitat). Rules of behavior for the southern pine beetle were used to define habitat suitability (KEL image).

RCW - Functional Heterogeneity

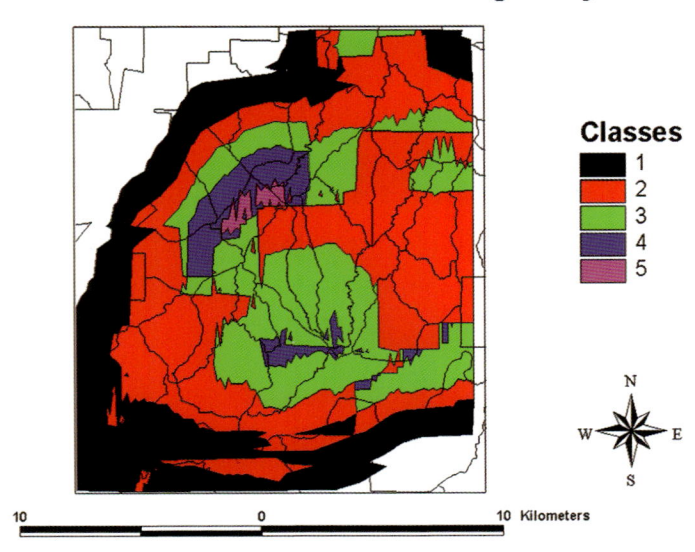

Figure 2.16 — "Functional heterogeneity" map illustrating habitat preference for the Red-cockaded woodpecker (1=least suitable habitat, 5=highly suitable habitat). Rules of behavior for the Red-cockaded woodpecker were used to define habitat suitability (KEL image).

**Combined Functional Heterogeneity Map
SPB and RCW**

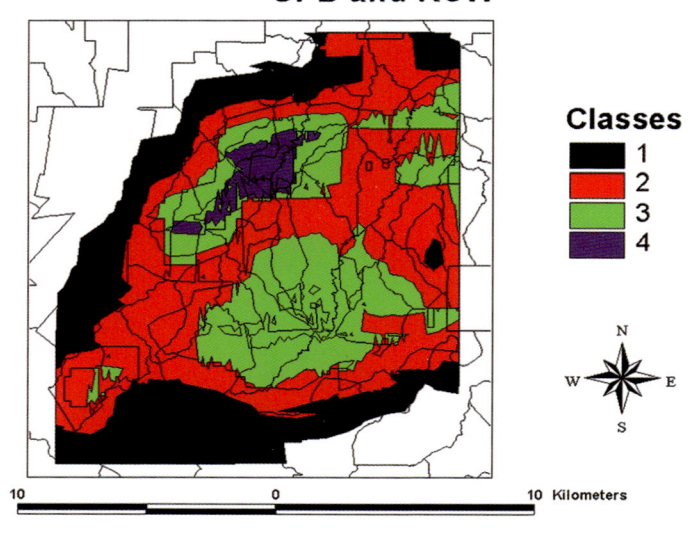

Figure 2.17 — The area of interaction of the southern pine beetle and the Red-cockaded woodpecker defined by overlaying the "functional heterogeneity" maps for the two species (1=least overlap, 4=high overlap). The southern pine beetle and Red-cockaded woodpecker utilize the same habitat within the forest landscape but for different reasons (KEL image).

selects landscapes containing longleaf pine for nesting, roosting, and foraging. Prior to European settlement, longleaf pine was a predominant species in the South, and it is likely that the southern pine beetle and Red-cockaded woodpecker rarely interacted, i.e., they were spatially separated by habitat preferences. This example illustrates an agenda item from community ecology, namely competition for habitat resources by two species populations.

Disturbance Effects on Forests – Ecosystem

Insect herbivores are common in forest ecosystems. Representatives of this functional group are so numerous that they are often grouped according to feeding preference, e.g., defoliating, sapsucking, terminal feeding, root feeding, seed and cone infesting, phloem boring, wood boring, and gall making (Coulson and Witter 1984). From a human perspective, the activities of herbivorous insects are manifold: They are selective mortality agents to forest trees, they alter plant community species composition and age structure, they weaken trees and increase vulnerability to plant pathogens and natural disturbances, they modify the growth form and appearance of trees, they reduce food supplies used by other herbivores, and they can enhance or reduce forest regeneration. Typically the activities of herbivorous insects go unnoticed (except by forest entomologists) until there is an epizootic event where elevated population levels result in impacts that exceed normal or expected levels. The term *impact* is defined broadly to mean any effect on the forest ecosystem resulting from the activities of insects. The effects are seen as qualitative or quantitative change in conditions and/or resources. These epizootic events are viewed by ecologists as disturbances to the forest ecosystem. A *disturbance* is an event that results in a deviation or change from expected behavior of an ecosystem. This definition requires a reference state or condition, i.e., a normal or nominal state. So, for example, if herbivory (consumption) in a forest ecosystem is normally 20 percent of leaf biomass each year, defoliation that removed 50 percent or more would be considered a disturbance. Valuing impact of insect-initiated disturbances traditionally began with assessing economic loss of forest resources. This parochial view changed when Mattson and Addy (1975) suggested that phytophagous insects played a significant role in regulating primary production in forest ecosystems. Research directed to understanding insect-initiated disturbance effects on basic ecosystem processes continues (Coulson and Stephen 2006, Raffa et al. 2008).

The Coweeta Hydrological Laboratory in western North Carolina, administered by the USDA Forest Service, Southern Research Station, is a unique setting for the scientific study of ecosystem processes. The spatial extent of the Coweeta Basin is 2185 ha and it is situated within the Blue Ridge geologic province of Noth Carolina. Elevations range from 679 to 1592 meters. More than 50 kilometers of streams drain the basin. There are four major forest types present: northern hardwoods, cove hardwoods, oak-chestnut, and oak-pine. The Coweeta Basin is a quintessential example of a landscape, as per our definition: i.e., a spatially explicit geographic area consisting of recognizable and characteristic component ecosystems (Figure 2.18). The individual *ecosystems* that collectively constitute the landscape are represented as discrete watersheds (Figure 2.19). Wiers (small overflow-type dams) (Figure 2.20) were constructed on the watersheds to gauge hydrological events, and later some were equipped with instrumentation for stream chemistry measurements as well. One of the important features of the research conducted at the Coweeta Hydrological Laboratory is that the boundaries for studies are delineated by watersheds, i.e., the spatial scale (extent) of investigations is set by natural landform features. Interaction with adjacent ecosystems is thereby minimized; and this circumstance facilitates studies of basic hydrology and measurement of nutrient input, output, and internal cycling. It is also possible to manipulate the ecosystem processes through various types of treatments, and compare among different watersheds. Examples of past treatments include varying intensities of forest harvesting, conversion of hardwoods to grass, mountain farming, and application of herbicides and fertilizers.

In one of the first studies undertaken to investigate the ecological impact of insect herbivory in ecosystems, Swank et al. (1981) examined the effects of defoliation by the fall cankerworm, *Alsophilia pometaria* (Figure 2.21), on

Figure 2.18 — The Coweeta Basin, site of the Coweeta Hydrologic Laboratory in western North Carolina. This facility is administered by the USDA Forest Service, Southern Research Station and is also a Long Term Ecological Research (LTER) site. The Coweeta Basin is a quintessential example of a landscape, as per our definition: a spatially explicit geographic area consisting of recognizable and characteristic component ecosystems (KEL photograph).

primary production and nutrient cycling. The study was conducted at the Coweeta Hydrological Laboratory. The fall cankerworm forages many hardwood trees including oak, hickory, hackberry, boxelder, basswood, elm, apple, and maple. Feeding takes place in the spring and early summer (late April through June in western North Carolina). Outbreaks are common and generally last several years before populations collapse to enzootic levels (Coulson and Witter 1984). In 1969 epizootic populations occurred within the Coweeta Basin on Watershed 27, a 38.8 ha control catchment vegetated with undisturbed mixed hardwood species (Swank et al. 1981). The outbreak subsequently spread to Watershed 36, a 49 ha catchment, also vegetated with mixed hardwoods species. Populations collapsed in 1979. This herbivory-triggered disturbance had a measurable effect on nutrient export from the infested watersheds in comparison to a control watershed (Watershed 18, a 13 ha control catchment, also vegetated with mixed hardwood species). See Figure 2.19 to visualize the spatial relation of Watersheds 27, 36, and 18. Measurement of nitrate nitrogen (NO_3-N) in the streams of each watershed was used to evaluate the disturbance. The effects of defoliation on annual weighted concentrations and total export of NO_3-N on Watersheds 27 and 36 were typically four to five fold greater than observed on the undisturbed control catchment (Watershed 18). These results clearly demonstrated a functional ecosystem-level consequence of the feeding activities of the fall cankerworm. The significance of the increased loss of NO_3-N was in the alteration of nutrient transfer and turnover rates associated with forest defoliation, and the consequences of such functional changes for the persistence and metabolism of the forest ecosystems over long time periods (Swank et al. 1981).

Accelerated transport of NO_3-N from the defoliated watersheds was clearly demonstrated in Swank et al. (1981), but what was the impact of the disturbance within these ecosystems? Herbivore consumption of foliage is noteworthy for its inefficiency, and a great deal of the plant biomass is simply reconstituted as frass (insect feces) and returned to the forest floor. An indelible memory of ecologists who have worked in forests being defoliated by epizootic populations of insect herbivores is the sight and sound of a "frass shower." Does the disturbance act as a form of "forest fertilization"? Certainly consumption breaks down (fragments) and concentrates nutrients associated with leaves. Can decomposer organisms mobilize nutrients bundled as frass quicker than leaf litter? Is there a subsequence pulse in forest growth following defoliation? Does the loss of photosynthetic activity, as a consequence of the spring defoliation, cancel the effect of fertilization? If the herbivore activity occurred later in the summer, would plant growth be accelerated? These questions, and many others, are the substance of landscape ecological research targeted to ecosystems.

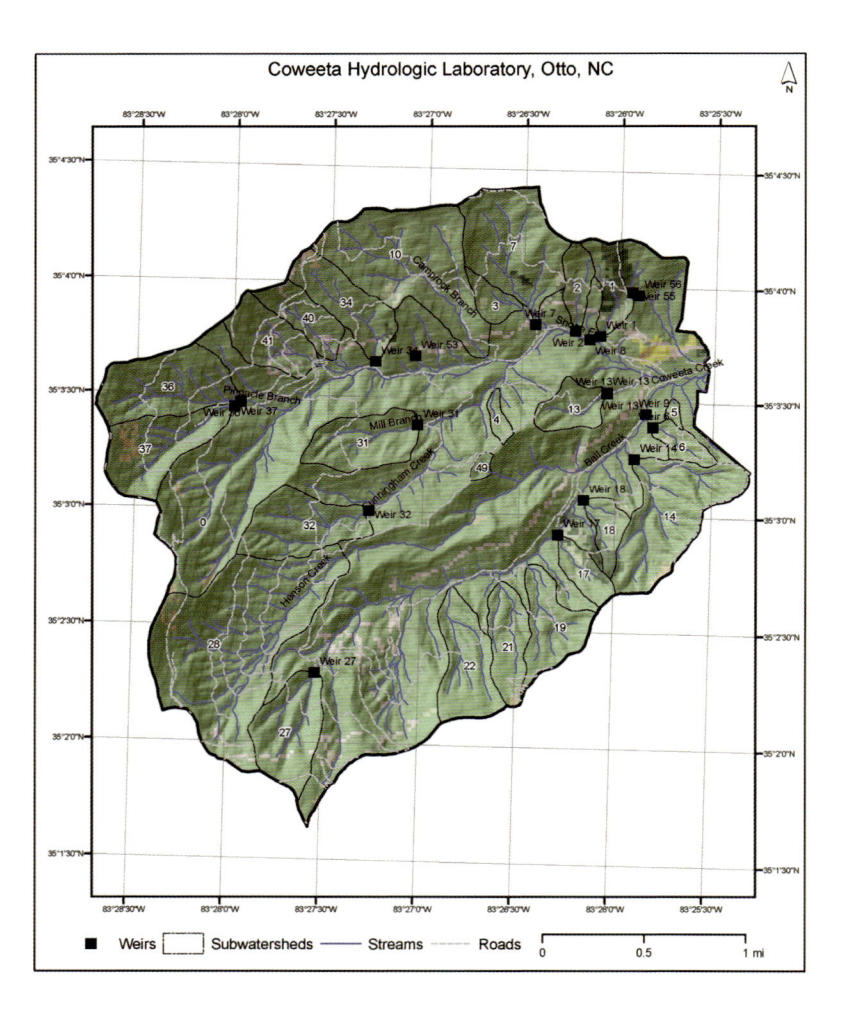

Figure 2.19 — Site map of the Coweeta Hydrologic Laboratory. The individual ecosystems that collectively constitute this landscape are represented as discrete watersheds (USDA, Forest Service map).

Figure 2.20 — Wiers, small overflow-type dams, were constructed at the base of the individual watersheds (catchments) at the Coweeta Hydrologic Laboratory to gauge hydrological events and facilitate stream chemistry measurements (USDA, Forest Service photograph).

Figure 2.21 — The fall cankerworm, *Alsophilia pometaria*. Epizootic population levels of this defoliating insect resulted in measurable changes in primary production and nutrient cycling within infested watersheds at the Coweeta Hydrologic Laboratory (photograph by Tim Tigner, VA Division of Forestry).

Organization of the Knowledge Base for Ecology | 33

EPILOGUE TO ORGANIZING THE DISCUSSION OF ECOLOGY

The concept of levels of ecological integration places landscape ecology in the context of the parent discipline, ecology. The concept of hierarchy provides a means of illustrating the relation of landscapes, ecosystems, and ecoregions (biomes). The concept of emergent properties illustrates the uniqueness of landscapes as an academic discipline. These concepts were introduced to simplify the organization of our discussion of landscape ecology. The examples illustrate how an agenda topic from population ecology, community ecology, and ecosystem ecology becomes a landscape ecological issue when examined in a spatial context.

3

Spatial and Temporal Scale in Landscape Ecology

OVERVIEW

The concept of scale is fundamentally important in the study of landscape ecology. The term *scale*, like landscape, has been claimed by various disciplines and, again, specialized uses and definitions exist. For our purposes there are three concepts of scale relevant to landscape ecology. The first concept deals with spatial and temporal extent. Typically, the initial task in describing a landscape involves definition of size, i.e., how large or small is the area of interest. Also, the ecological processes that operate within landscapes, which are often the focus of ecological study, are time-dependent, and knowledge of temporal duration is therefore important. The second concept of scale deals with maps and images that together represent the fundamental media of landscape ecology. There are basic rules that govern the use of maps and images for representing and comparing spatial information. The third concept of scale deals with quantitative ecology. Measurements in ecological studies are defined by units and dimensions. Procedures for numerical analysis and synthesis of ecological information are based on measurements expressed as units and dimensions. Below we examine and illustrate each concept (Figure 3.1).

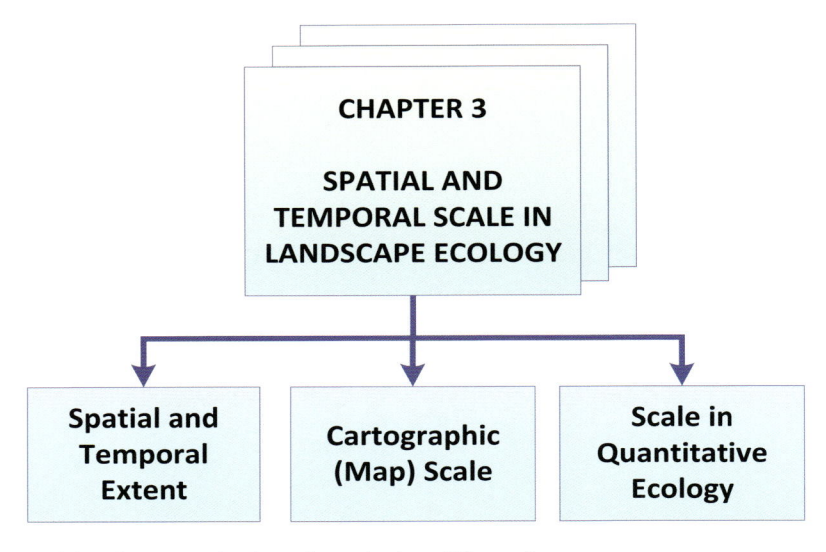

Figure 3.1 — Summary of topics and organization of Chapter 3.

SPATIAL AND TEMPORAL EXTENT

Spatial extent (scale) is simply the size of an area of interest. Size is usually expressed as m, m^2, or m^3. *Temporal extent* (scale) is the length of time for an observation or an event to take place. Time is usually expressed in chronological units (seconds, hours, days, months, years) or occasionally in relation to events (daily, lunar, annual cycles). Spatial and temporal extent are often categorized by descriptors such as micro-, meso-, or mega-; large- or small-; broad- or narrow-; fine- or coarse-. Typical boundaries for space and time are defined in Table 3.1.

Note that in the context of spatial and temporal extent, these descriptors (Table 3.1) generally have the normal or expected meaning, e.g., a large scale study would include a large area. However, the dimensions of these descriptors have not been uniformly applied in the ecological literature. See Bailey (1996) for alternative specifications. In particular, the temporal scale in ecological studies is often much more towards the bottom end of the microscale bracket than the top end in Table 3.1. For this reason, sometimes a distinction is made between "ecological" time vs. "geological" time, the implication being that ecological time is shorter than geological time (Figure 3.2).

Table 3.1 — Space/time domains (Delcourt and Delcourt 1992).

Scale	Time
Microscale	1 to 500 years
Mesoscale	500 to 10,000 years
Macroscale	10,000 to 1,000,000 years
Megascale	1 million to 4.6 billion years

Scale	Size
Microscale	1 m^2 to 100 ha
Mesoscale	100 to 1 000 000 ha or 1000 km^2
Macroscale	1000 to 1 000 000 km^2
Megascale	Over 1 000 000 km^2

Many studies and applications in landscape ecology involve the use of digital maps and images. In order to quantitatively describe, analyze, and compare information about landscapes captured on digital maps and images, another dimension of spatial scale is needed. In addition to spatial extent (range), we need to add a measure of resolution (grain). The term *resolution* (grain) is equivalent in concept to pixel (short for *pic*ture *el*ement). Figure 3.3a illustrates the affect of progressively increasing the resolution (by decreasing the grain size) in a sequence of images of an agricultural landscape. In Figure 3.3b the resolution is held constant and the range is progressively increased. Both the range (spatial extent) and resolution (grain or pixel size) in a digital image or map are variables that can be adjusted. One of the challenges of research is to define the proper or appropriate spatial scale necessary for a specific landscape ecological study. Of course, the researcher can experiment by varying the spatial scale.

Once the range and resolution have been defined, the next step in using a digital image or map is to identify (classify) the features of interest. For example, in Figure 3.4, vegetation cover is the variable of interest, i.e., the plant communities associated with the landscape. To classify the landscape, each plant community would be assigned a unique number. By inspecting an image of the landscape, the ecologist could identify the different plant communities, locate where the communities occur on the

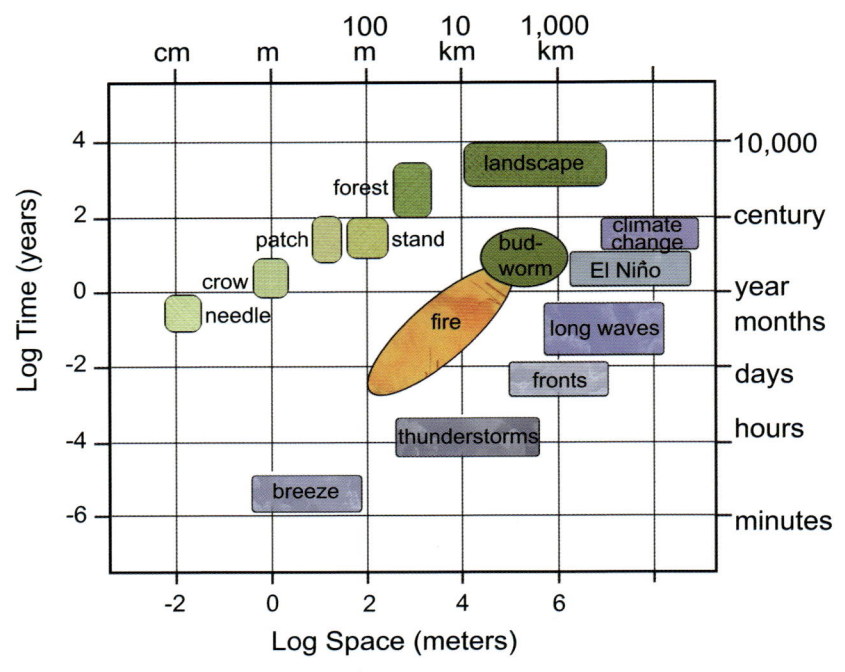

Figure 3.2 — Temporal and spatial scales of the boreal forest (Holling 1986), of the atmosphere (Clark 1985), and of their relationships to some of the processes that structure the forest. "Contagious meso-scale processes such as insect outbreaks (budworm) and fire mediate the interaction between faster atmospheric processes and slower vegetative processes" (modified from Gunderson and Holling 2002).

image, and assign a value to each cell (pixel). Figure 3.5 illustrates the plant communities associated with a coastal prairie landscape in South Texas that was the site of a feral honey bee (*Apis mellifera*) study.

In landscape ecological studies, specification of the appropriate temporal scale is also of fundamental importance. The concept of time parceled out in chronological units is a human device dating from the 14th century. Before then, time was measured by the occurrence of natural events in daily, lunar, and annual cycles. Consideration of time as event-driven is a useful aid in defining an appropriate temporal scale for a landscape ecological study (Coulson et al. 1987). In effect, the question is: How does the organism of interest in an ecological study perceive and respond to time?

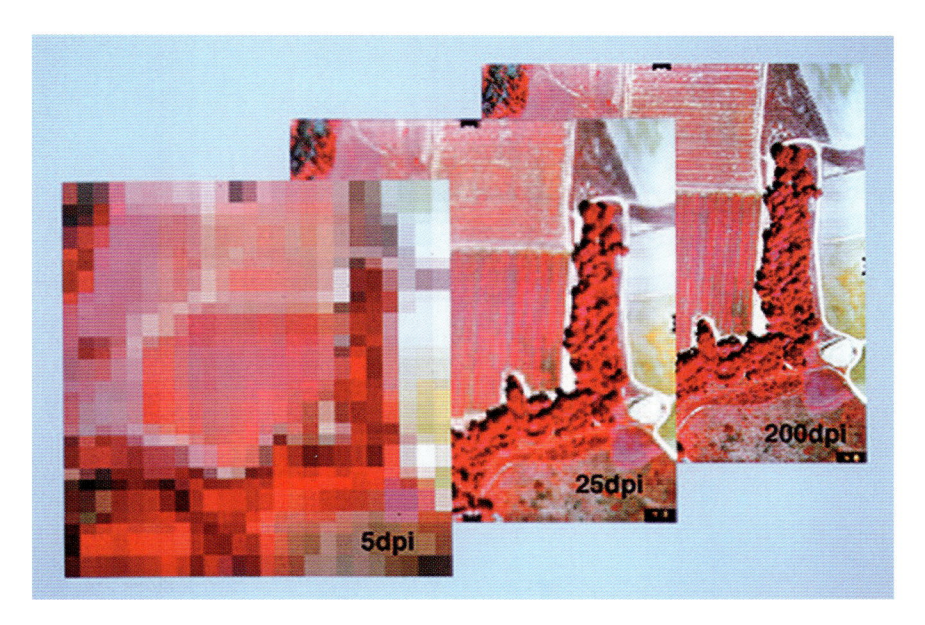

Figure 3.3a — Aerial view of an agricultural landscape. In this image the range (spatial extent) is held constant and the resolution is increased. In panel (1) the resolution is five pixels (cells) per inch, in panel (2) the resolution is 25 pixels per inch, and in panel (3) the resolution is 200 pixels per inch (KEL image).

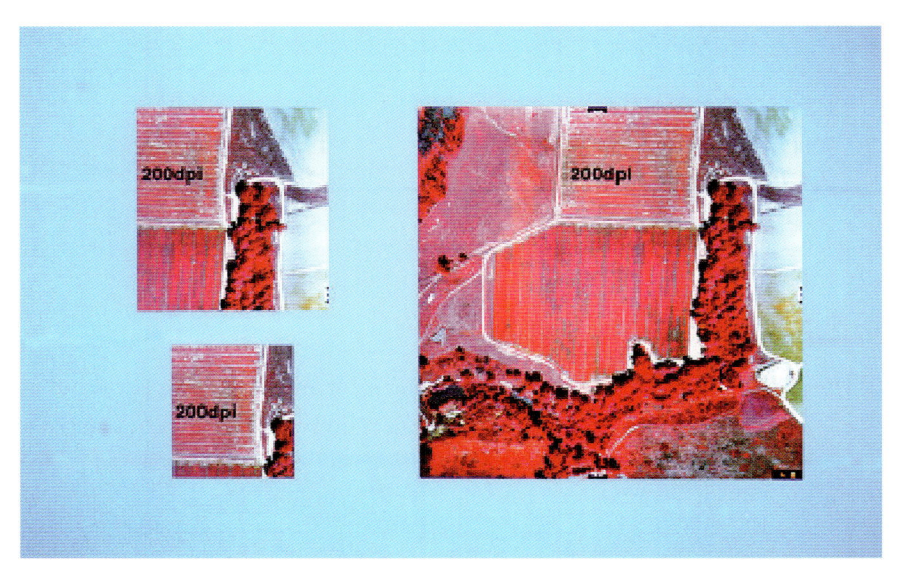

Figure 3.3.b — Aerial view of an agricultural landscape. In this image the resolution is held constant and the range (spatial extent) is increased progressively (KEL image).

Spatial and Temporal Scale in Landscape Ecology | 39

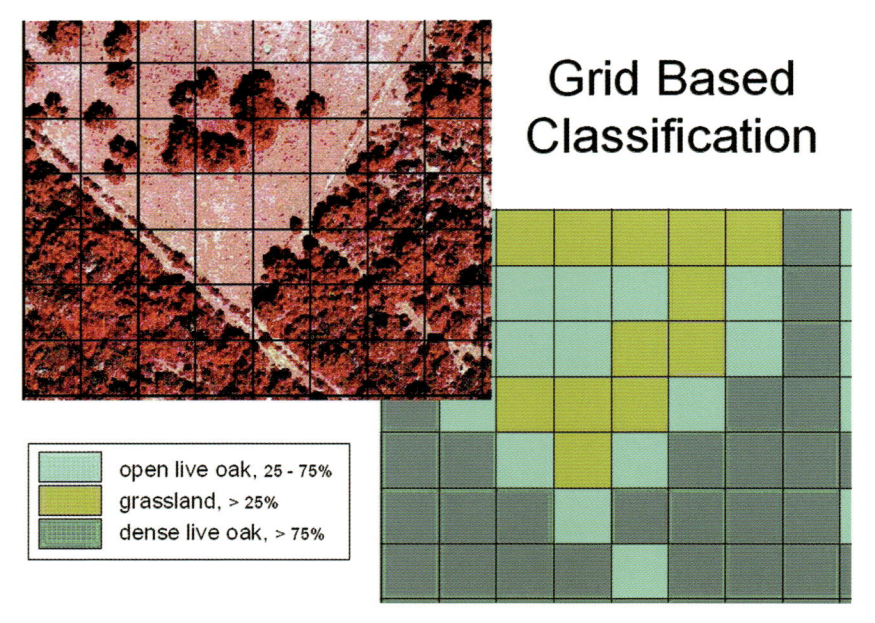

Figure 3.4 — Once the range and resolution have been defined, the next step in using a digital image or map is to identify (classify) the features of interest. In this image, vegetation cover is the variable of interest, i.e., the species of plants associated with the landscape. To classify the landscape each plant community would be assigned a unique number. By inspecting an image of the landscape, the ecologist could identify the different plant communities, locate where the communities occur on the image, and assign a value to each cell (pixel) (KEL image).

CARTOGRAPHIC (MAP) SCALE

Cartography is the art, science, and technology of making maps. A *map* is a graphic representation of the cultural and physical environment. Most maps are reduced representations, i.e., they are physically smaller than the real world.

In cartography, *scale* is defined as the ratio between the size of an area on a map and the actual size of that area on the earth's surface. Scale, which is a requisite feature of all maps, can be communicated in several ways, e.g., verbal expression, a scaled line, pictorial symbol, labeled feature, and a representative fraction (Gersmehl 1996) (Figure 3.6).

The representative fraction is a common way of depicting map scale. A *representative fraction* (RF) scale defines how many units of real-world distance are represented by one unit on the map. An RF scale of 1:1,000

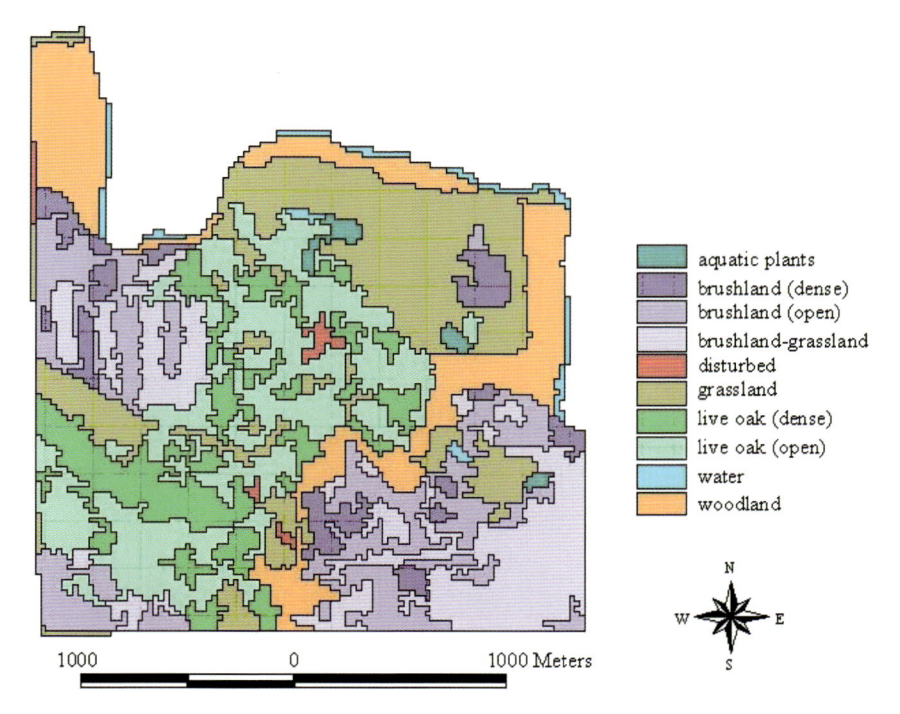

Figure 3.5 — Classified image of a coastal prairie landscape in South Texas illustrating plant community composition. This landscape was used in a study of feral honey bees (KEL image).

means that one unit of measurement on the map represents 1,000 of the *same* units in the real world. If the scale on a map is 1:24,000, then 1 inch on the map represents 24,000 inches, or 2,000 feet on the ground (24,000 inches divided by 12 inches = 2,000 feet). A small scale [small fraction] (e.g., 1:250,000) represents a large area on the ground. By contrast a large scale [large fraction] (e.g., 1:24,000) represents a small area on the ground. This example illustrates the essence of the problem with the cartographic definition of scale in reference to landscape ecological applications: The *smaller the scale* of the map, the larger is the area it covers and the more generalized are the data it portrays; the *larger the scale*, the smaller is the area depicted and the more specific are the data it portrays. Large-scale maps show small areas, and small-scale maps show larger areas (Figure 3.6). Much of the United States has been mapped by the U.S. Geological Survey (USGS) at a scale of 1:24,000. USGS maps at this scale cover an area measuring 7.5 minutes of latitude and 7.5 minutes of longitude. These

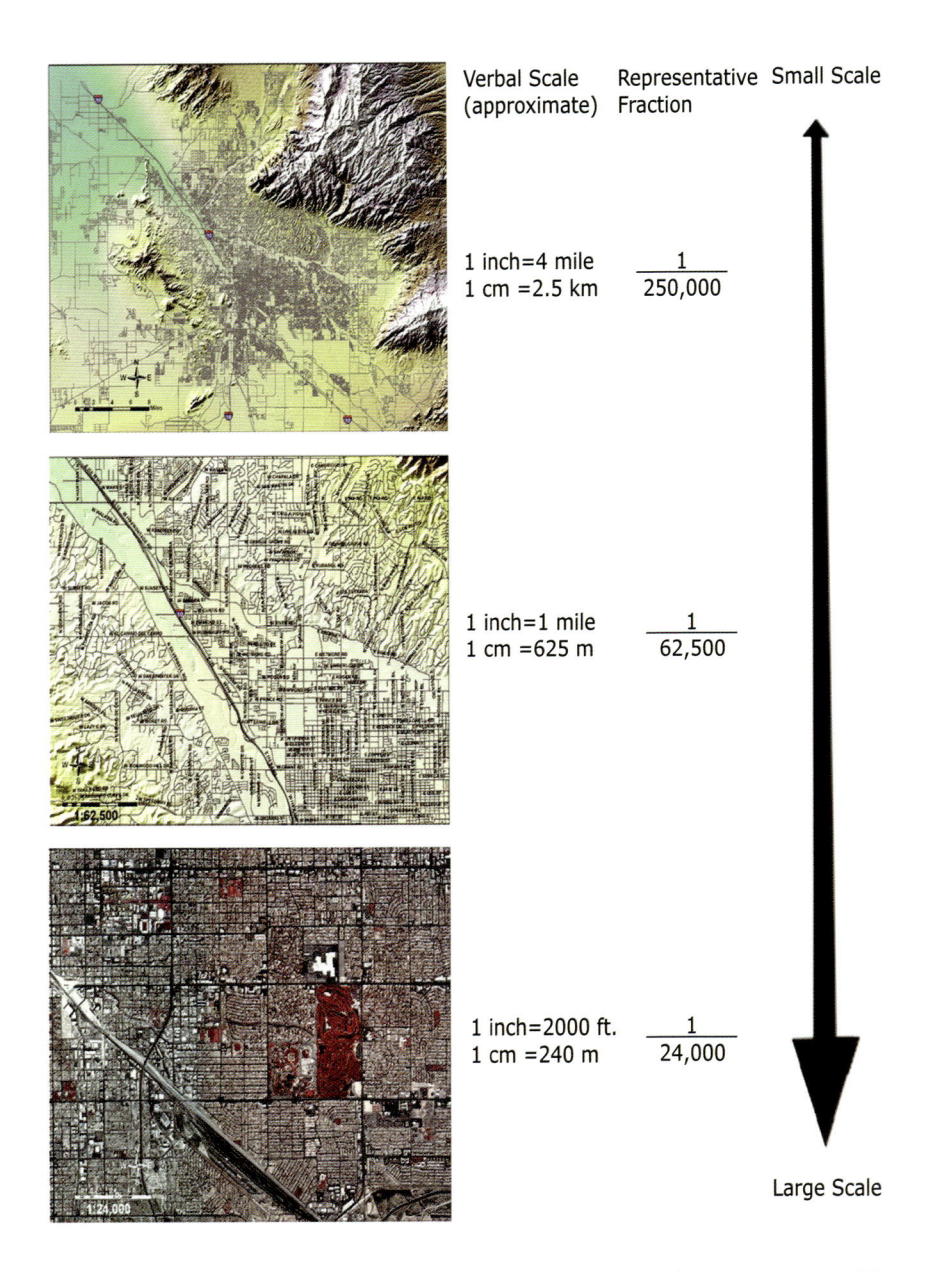

	Verbal Scale (approximate)	Representative Fraction	
	1 inch=4 mile 1 cm =2.5 km	$\frac{1}{250{,}000}$	Small Scale
	1 inch=1 mile 1 cm =625 m	$\frac{1}{62{,}500}$	
	1 inch=2000 ft. 1 cm =240 m	$\frac{1}{24{,}000}$	Large Scale

Figure 3.6 — In cartography, scale is defined as the ratio between the size of an area on a map and the actual size of that area on the earth's surface. Scale, which is a requisite feature of all maps, can be communicated in several ways, e.g., verbal expression, a scaled line, pictorial symbol, labeled feature, and a representative fraction (KEL image).

maps are commonly referred to as 7.5-minute quadrangle maps <http://www.usgs.gov>.

Digital images and maps are an integral part of the study of landscape ecology. The remarkable developments in geographic information system (GIS) technologies have greatly facilitated the use of maps and images in landscape ecology. However, the landscape ecologist still needs to understand basic concepts of spatial scale, as well as coordinate systems and map projections. These topics are well-worked in an introductory GIS course.

SCALE IN QUANTITATIVE ECOLOGY

Ecology is a science where the objects of interest or study (plants, animals, the elements of the environment) can be described by *units* and *dimensions*. Although these two terms are part of our common parlance, they have precise definitions when used in science. *Metrology* is the field of study that addresses measurement. In the United States, the National Institute of Standards and Technology (NIST), which is part of the Department of Commerce, is the gatekeeper for the International Systems of Units, or *SI*.

There are seven *SI* units that are defined by the metric system. The *SI* units include the following: mass, length, temperature, amount of a substance, time, electric current, and luminous intensity. The appropriate units and abbreviations for the *SI* units are defined in Table 3.2. The basic *SI* units can be combined to provide *derived SI* units, the most important of which for most ecological applications are *area* (length x length [m^2]) and volume (area x length [m^3]). These derived *SI* units provide the units and dimensions for ecological studies. For example *biomass,* the total mass of all living organisms per unit of surface area, is expressed as g/m^2. Furthermore, the patterns and processes associated with the operation of ecological systems, which are of great interest to ecologists, are also described in terms of units and dimensions. *Productivity* of a forest could be expressed as the change in biomass per unit of surface area per unit of time: $g/m^2/yr$. Ecology deals in part with the space, time, and mass components of quantities (Schneider 1994). For this reason an understanding of the quantitative concept of scale is extremely important as it is the foundation for analysis of data associated with ecological studies.

Table 3.2 — The basic International Systems of units, *SI* (National Institute of Standards and Technology, US Dept. of Commerce).

Physical Quantity	Name of Unit	Abbreviation
Mass	kilogram	kg
Length	meter	m
Temperature	Kelvin	K
Amount of substance	mole	mol
Time	second	s
Electric current	ampere	A
Luminous intensity	candela	cd

The quantitative ecological view of scale has several implications to the study of landscape ecology. First, ecological patterns emerge from the analysis of data at characteristic space and time scales (Wiens 1989). Second, measurements that define ecological patterns and processes occur within a characteristic range of values. Third, the analysis of scaled quantities defined by units and dimensions provides the means for understanding patterns and processes (Schneider 1994). And, finally, patterns and processes associated with populations, communities, ecosystems, and landscapes can be investigated across a range of spatial and temporal scales (Figure 3.2). Recall that our definition of landscape ecology was the science that dealt with the agenda of ecology in a spatial context.

EPILOGUE TO THE DISCUSSION OF SCALE

We examined three different concepts of scale fundamental to our study of landscape ecology. The first dealt with spatial and temporal extent. The second dealt with the cartographic representation of space. The third dealt with the use of scale in quantitative ecology. In Chapter 2 we introduced the concepts of levels of ecological integration and hierarchy as a means of organizing the discussion of the science of ecology. Remember that the levels of integration are simply compartments containing specific information, concepts, and principles that collectively represent the knowledge base for a subject of interest (Figure 2.2). The levels in the hierarchy that represent our focus on ecological science (populations, communities, ecosystems, and landscapes) do not have a characteristic or representative scale. We have indicated that ecosystems (ecotopes) and landscapes can be defined specifically by geographic coordinates, but there is not a specific range or resolution for either. That is to say there is not a characteristic spatial or temporal scale associated with each of the levels of integration. Ecological patterns and processes associated with a specific level of integration (e.g., the population level of integration) can be examined at different temporal and spatial resolutions and ranges. In fact, the need for multi-scale analysis is an important conclusion that has emerged from the various examinations of the concept of scale in ecology (Wiens 1989, Levin 1992, Schneider 1994). It is also noteworthy that concepts of scale play a fundamental role in how we envision and interpret information for purposes other than studying landscape ecology (Tufte 1983, 1990, 1997, 2006).

4

The Ecosystem Concept in Landscape Ecology

OVERVIEW

In Chapter 2, we introduced definitions of *ecosystem, landscape,* and *ecoregion* (biome). Each of these levels of ecological integration, in addition to having a substantial knowledge base associated with it, is founded on a conceptual underpinning that extends substantially beyond the simple definitions given. Focus in this chapter is directed to the ecosystem level of integration. A. G. Tansley (1935), a plant geographer, is credited with coining the word *ecosystem*. E. P. Odum (1953) articulated the ecosystem concept and placed it into the mainstream of ecology. F. G. Golley (1993) served as historian and traced the evolution and application of the concept. Today, the term *ecosystem* pervades both scientific discussion in ecology and popular commentary on the environment. Consequently, there are several different, but correct, views of what constitutes an ecosystem (Pickett and Cadenasso 2002)[1].

Our ecological definition of landscape is based on the concept of ecosystem, i.e., a spatially explicit geographic area consisting of recognizable and characteristic component *ecosystems*. If this approach to the presentation of landscape ecology is to be meaningful, a basic understanding of the

---

[1] To verify this statement, define your view of the term *ecosystem* and see how it matches or contrasts with the commentary that follows.

ecosystem concept is essential. Accordingly, the purpose of this chapter is to examine fundamental elements of the ecosystem concept. Emphasis is placed on the ecological scientific concept and we consider issues associated with structure, function, and change in ecosystems. We also discuss how the concept is applied to describe the environment around an organism or geographic area of interest and how it is used as a metaphor for holism or systems thinking. We conclude by considering two anthropocentric concepts that apply knowledge of ecosystems structure, function, and change, i.e., ecosystem services and ecosystem management (Figure 4.1).

ECOSYSTEM STRUCTURE

An *ecosystem* is defined simply as the biotic community and the abiotic environment functioning together. Since first introduced this definition has been elaborated upon, and the following statement captures the essence of the concept:

"Any unit (a biosystem) that includes all the organisms that function together (the biotic community) in a given area interacting with the physical environment so that a flow of energy leads to clearly defined biotic structure and cycling of materials between living and nonliving parts is an ecological system or *ecosystem*" (Odum 1983).

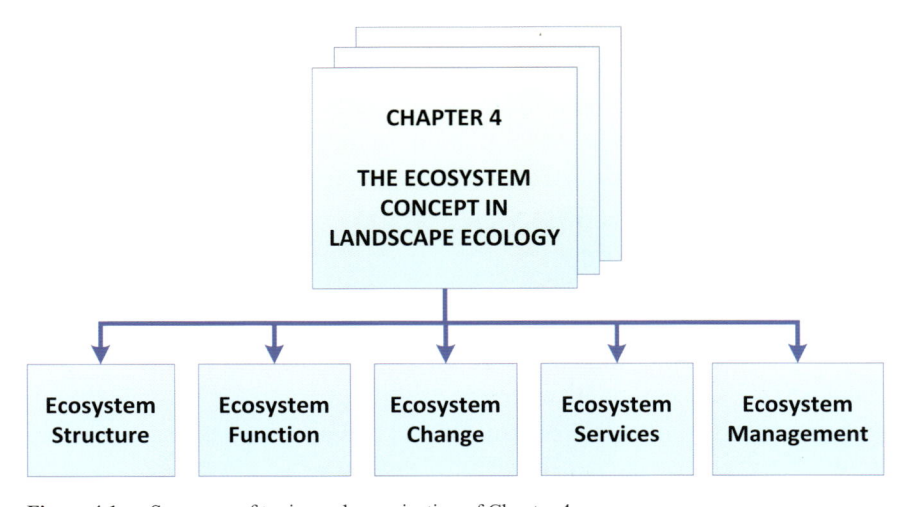

Figure 4.1 — Summary of topics and organization of Chapter 4.

Professor Odum's concept places emphasis on both structure (components) as well as function (the flux of energy, materials, and information), but is inherently vague with regard to space and time. A logical question that follows from the definition is: How large or small is an ecosystem? An alternative definition that begins to address this question is as follows: An *ecosystem* is "...a spatially explicit unit of the Earth that includes all of the organisms, along with all components of the abiotic environment within its boundaries" (Likens 1992). By including boundary, this second definition addresses (in part) the issue of how large or small an ecosystem is, i.e., its spatial extent (Figure 4.2). This definition of ecosystem is equivalent in meaning to *ecotope* (proposed by Troll (1963), cited and discussed in Zonneveld (1995)), i.e., an *ecotope* is a bounded ecosystem; and also is synonymous with *site*, proposed by Bailey (1996).

However, there are additional ways the term *ecosystem* is applied. For example, ecosystem is often used to describe the environment around a special organism of interest, e.g., the forest ecosystem, the cotton ecosystem, the mangrove ecosystem (Figure 4.3). A forester, agronomist or entomologist might describe the domain of their research interest using this expression. Further, the term is frequently given to describe land and aquatic areas of interest, e.g., the Yellowstone Ecosystem, the Everglades Ecosystem, the Chesapeake Bay Ecosystem (Figure 4.4) (Coulson et al. 1999). A land- or water-use manager might describe the geographic area of their charge in this manner. In both of these cases the term *ecosystem* is a metaphor for holism or systems thinking and includes both scientific as well as social contexts. These uses of the term *ecosystem* are both common and correct.

ECOSYSTEM FUNCTION

All ecosystems have four basic processes: primary production, consumption, decomposition, and abiotic storage. *Ecosystem function* deals with the operation of these processes over time. Information about ecosystems can be organized using a simple model (Figure 4.5) that illustrates the fundamental constituents, their relationship, and pathways for materials cycling and energy flow (Crossley et al. 1984). Following is a brief description of each process.

Figure 4.2 — Color infrared photograph of a complex landscape consisting of multiple ecosystems. The individual polygons forming the mosaic represent "bounded ecosystems." How many different types of ecosystems can you identify? See Figure 5.2a and b for the answer. There are different types of agriculture, lakes, forests, grasslands, etc. Photograph from a post oak-savanna landscape in Central Texas (KEL image).

In terrestrial and aquatic ecosystems, *primary production* (a quantity) and *primary productivity* (a rate) are terms used to describe the conversion of solar energy by green plants via photosynthesis to organic substances. Gross primary production is the total organic matter converted, including that used by the plant, while net primary production is the amount stored in a plant in excess of its respiratory needs (Odum 1983). Green plants are referred to as *autotrophic* organisms or primary producers. *Consumption* is the total intake of food or energy by an organism during a specified period of time. Consumer organisms are referred to as *heterotrophs*. If the heterotroph consumes plant material it is a *herbivore,* and if it consumes other animals it is referred to as a *carnivore* (e.g., insect parasitoids and predators). When organic compounds cease to be components of a living system after death of the organism or excretion, they undergo a process

Figure 4.3 — The term *ecosystem* is often used to describe the environment around a special organism of interest. In this example the "organism of interest" is a mangrove forest (KEL image).

Figure 4.4 — The term *ecosystem* is frequently given to describe land and aquatic areas of interest, e.g., the Yellowstone Ecosystem, the Everglades Ecosystem, the Chesapeake Bay Ecosystem. In this example the "area of interest" is the everglades, located in South Florida (NPS photograph).

of alteration and adjustment known as decomposition. *Decomposition* is the breakdown of complex energy-rich organic molecules into simple inorganic constituents. The process involves the loss of heat energy and the conversion of organic nutrients into inorganic ones. In terrestrial ecosystems the process takes place largely on or within the soil. The organisms involved in the process are referred to as *detritivores* and most of the animal species belong to the Arthropoda. Decomposition involves the fragmentation (mechanical) breakdown of the litter and subsequent alteration of chemical structure (Coleman et al. 1996). The final ecosystem process is *abiotic storage*. It involves the interplay of weathering (the breakdown of rock materials by mechanical and chemical means) and decomposition on soil-forming processes. One result of this interaction is the assembly of pools of nutrients that can be used by primary producers for growth and development.

Nutrient cycling is the transformation of chemical elements from inorganic form in the environment to organic form in organisms and, via decomposition, back into inorganic form. Figure 4.5 illustrates pathways for nutrient cycling in ecosystems. Much of our knowledge of the functional roles of organisms in ecosystems came from investigations dealing with nutrient dynamics.

The behavior of energy in ecosystems is termed *energy flow* (Odum 1983). Ecosystems are open systems in which order is maintained by utilizing energy from the environment. As we have seen (Figure 4.5), the ultimate energy source in terrestrial and aquatic ecosystems is derived from the transformation of radiant solar energy by primary producers via photosynthesis. Energy flow in ecosystems is through the medium of living organisms. Flow is governed by the first and second laws of thermodynamics, i.e., (1) energy may be transformed from one type into another but is neither created nor destroyed (also known as the law of conservation of energy); and (2) the entropy of an isolated system not in equilibrium will tend to increase over time, approaching a maximum value at equilibrium (a measure of the disorder of a system). Pathways of energy flow are illustrated in Figure 4.5. Without constant energy input, ecosystems degenerate and eventually cease to function. You can verify this fact by placing your aquarium or terrarium in a room with no sunlight.

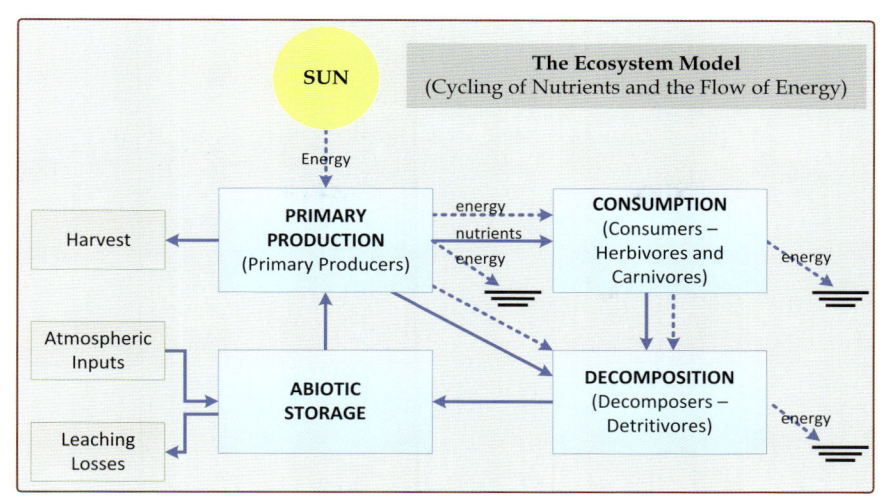

Figure 4.5 — Information about ecosystems can be organized using a simple model that illustrates the fundamental constituents (primary production, consumption, decomposition, and abiotic storage), their relationship, and pathways for materials cycling and energy flow (modified from Crossley et al. 1984).

The functional view of an ecosystem consisting of four basic processes (primary production, consumption, decomposition, and abiotic storage) linked by energy flow and nutrient cycling, introduces a new consideration for our discussion of ecosystem size. In most cases identification of whether or not the elements of function are present in an ecological unit (system) is straightforward (Figure 4.6, panels a and b). In other cases your image of an ecosystem will be challenged (Figure 4.6, panel c).

ECOSYSTEM CHANGE

Ecosystems change in space and time. The traditional view of ecological succession progressing in a linear fashion through an ordered series of stages beginning from a disturbed site and ending in a stable-state or climax community has undergone a substantial revision. The contemporary perspective of ecosystem change involves two provocative concepts: the adaptive cycle and panarchy (Gunderson and Holling 2002). Each concept is described below.

The first concept, the adaptive cycle, is a metaphor, a heuristic theory, of change in ecosystems. The process of change can be envisioned as

Figure 4.6a — Forest ecosystem located in the Smoky Mountains of western North Carolina. Can you envisage the four basic processes (primary production, consumption, decomposition, and abiotic storage) operating in this ecosystem? The gray skeleton trees are hemlocks killed by the hemlock woolly adelgid, *Adelges tsugae*, an herbivorous consumer organism (KEL image).

Figure 4.6b — Stream ecosystem, Slick Rock Creek, located in the Smoky Mountains of western North Carolina. Can you envisage the four basic processes operating in this ecosystem? (KEL image).

Figure 4.6c — Would the boulder in this stream be considered an ecosystem? Can you identify the four basic ecosystem processes? Primary producer organisms are clearly visible. Are the other processes in play? D. A. Crossley Jr., Professor Emeritus at the University of Georgia, challenged students in insect ecology with questions about the bounds of ecosystems (KEL image).

an *adaptive cycle* that represents ecosystem dynamics to consist of four distinct functional phases, referred to as: (1) growth or exploitation [**r**], (2) conservation [**K**], (3) collapse or release [**Ω**], and (4) reorganization [**α**]. These stages in the process roughly correspond to birth, growth, death, and renewal. The phases and their relationship are illustrated in Figure 4.7. Each preceding phase of the cycle creates the conditions needed for the next phase. This general model of systematic change in ecosystems results in an adaptive cycle with two major transitions. The first transition takes place from exploitation (**r**) to conservation (**K**) and is referred to as the forward loop. It is a slow, incremental, and predictable transition that results in growth and accumulation of *capital* (energy, materials, and information) within the ecosystem over time. The organization (*connectedness*) of the ecosystem increases during the transition. This forward loop encapsulates the traditional view of ecological succession. Stability and productivity of the ecosystem are determined in this sequence. The second transition takes place from release (**Ω**) to reorganization (**α**) and is referred to as the back loop. It is a fast and often unpredictable transition that leads to restructuring or destructuring of the ecosystem. The back loop is where adaptive change in the ecosystem can take place. *Resilience* (see definition on page 58) and recovery of the ecosystem are determined in this sequence. The changes taking place in this transition are facilitated by mutations; the occurrence of chance events; novel combination, recombination, or arrangement of elements; etc. In some instances the "experimentation" associated with the back loop results in a fundamentally different type of ecosystem. The transition from the forward to back loop of the adaptive cycle (**K→Ω**) is triggered by a variety of change agents, e.g., atmospheric-generated disturbances (hurricanes, wildfire, windstorms, ice storms, etc.), activities of invasive species (insect outbreaks, disease outbreaks), operation of the geomorphic processes (e.g., mass movement), and the activities of humans

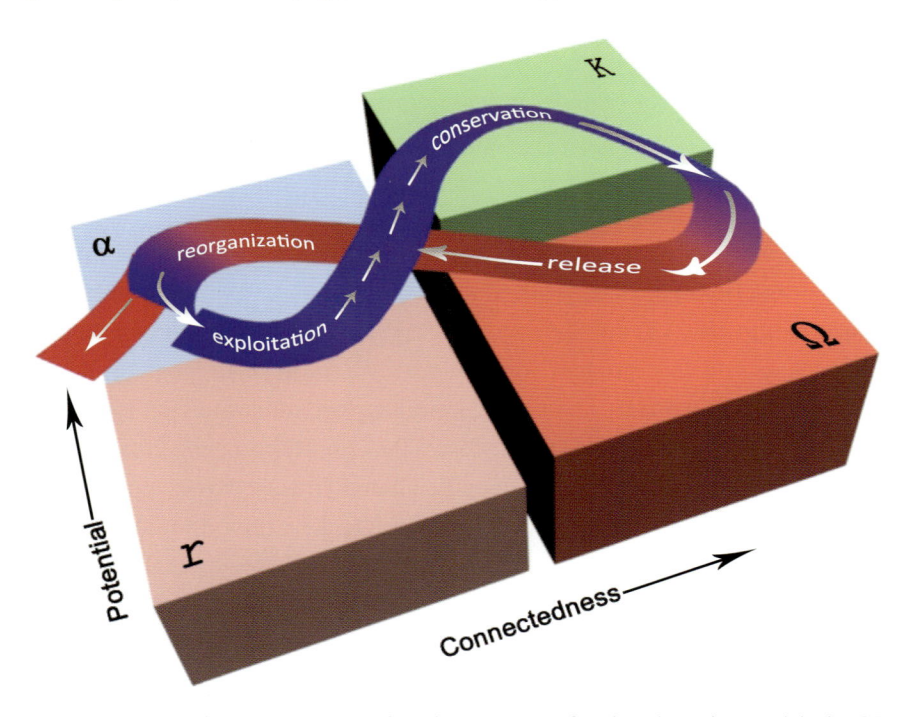

Figure 4.7 — A stylized representation of the four ecosystem functions (growth or exploitation [r], conservation [K], collapse or release [Ω], and reorganization [α]) and the flow of events among them. The arrows illustrate the speed of flow in the cycle: short and closely spaced arrows indicate a slowly changing situation and long arrows indicate a rapidly changing situation (modified from Gunderson and Holling 2002).

(e.g., through domestication of natural ecosystems). The transition from the back loop to the forward loop (α➔r) is the gateway to adaptive change in ecosystems. The mechanisms that facilitate this transition are poorly understood (Holling 1992, 2001, 2004; Gunderson et al. 1995; Gunderson and Holling 2002).

From this basic concept of the adaptive cycle, a second concept follows: a general theory of change in complex adaptive systems, referred to as *panarchy* (Gunderson et al. 1995, Holling 2001, Gunderson and Holling 2002). This theory considers ecosystems to consist of a nested set of four-phase interactive adaptive cycles that occur at different spatial and temporal scales (Figures 4.8 and 3.2). Taken together the set of interactive cycles is referred to as a *panarchy*. Two features distinguish panarchies for hierarchies. The first is the adaptive cycle *per se*, particularly the α phase which serves as the "engine of variety and the generator of new experiments" (Holling 2001) within each level in the panarchy. The second feature is the connection between levels in the panarchy. Multiple connections between phases at one level and phases at another level in the panarchy are possible.

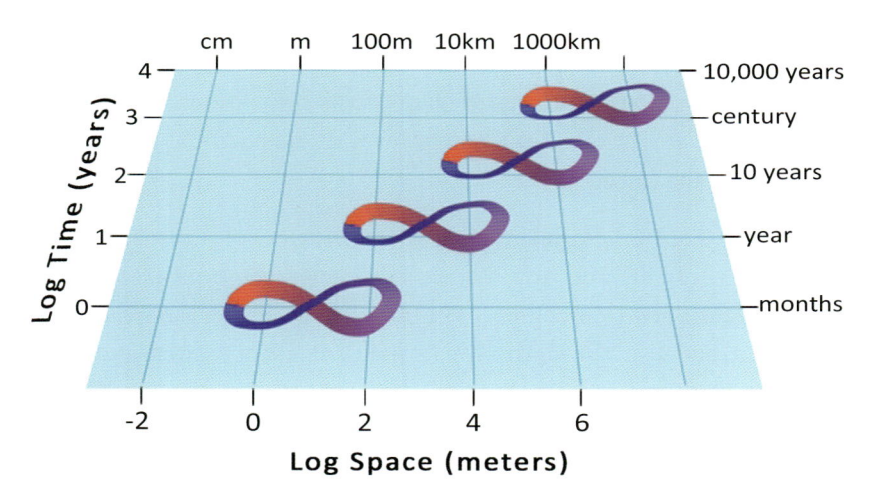

Figure 4.8—A stylized panarchy. A panarchy is a cross-scale, nested set of adaptive cycles, indicating the dynamic nature (space and time) of the various processes that structure ecosystems (see Figure 3.2 for examples of specific processes associated with a forest ecosystem). Note that we are using the term ecosystem in the holistic metaphor sense, i.e., a land area of interest (modified from Gunderson and Holling 2002).

The Ecosystem Concept in Landscape Ecology | 57

Two connections in the panarchy, described as *revolt* and *remember*, are particularly significant (Figure 4.9). Each is examined below:

Revolt is the situation where a disturbance event initiated in a fast and small cycle impacts slow and larger cycles. An example is a forest fire. Ignited by a lightning strike, a fire originating in an individual tree can spread to a patch in the forest and then to an entire stand of trees. In this example three separate levels in the panarchy are affected. Revolt initiates from the Ω phase of the fast and small cycle and cascades upward to the slow and larger cycles, having the greatest consequence when the affected cycles are in the **K** phase where resilience is low. *Resilience* in a panarchy is defined as the capacity of a system to experience disturbance and still maintain its ongoing functions and controls (Gunderson and Holling 2002).

Remember is the second type of cross-scale connection in a panarchy. This connection links a slow and large cycle to a fast and small cycle. Remember takes place when a level in the panarchy is in the indeterminate α phase and the configuration of the next higher level is in the **K** phase (Figure 4.9). The remember connection facilitates renewal by drawing on the *potential* [capital] (energy, materials, and information) that has been accumulated and stored in a larger, slower cycle. For example, following an insect outbreak that caused mortality to a patch of trees, the accumulated capital associated with the forest stand (the next higher level in the panarchy) preserves physical structure, entrains nutrients, provides propagules from the seed bank, etc. In effect the higher level of the panarchy contributes a "memory" to the lower level.

In summary, Figure 4.9 illustrates how small-fast cycles can affect larger slower ones (revolt) and how large-slow cycles can control renewal of smaller faster ones (remember). The fast levels in the panarchy invent, experiment, and test. The slower levels stabilize and conserve accumulated memory of past successful, surviving experiments. Taken as a whole, the panarchy is both creative and conserving. The interactions between cycles in a panarchy combine learning with continuity <http://www.resalliance.org/570.php> (Gunderson et al. 1995, Holling 2001, Gunderson and Holling 2002).

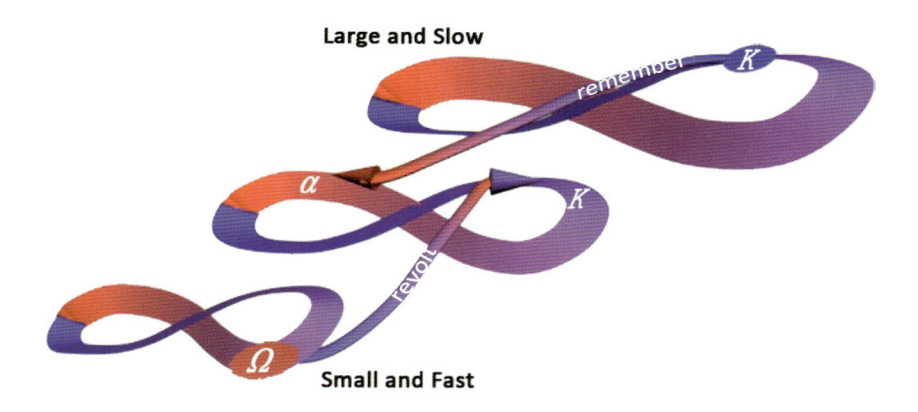

Figure 4.9 — Panarchical connections. To emphasize the two connections that are critical in creating and sustaining adaptive capability, three selected levels of a panarchy are illustrated. One connection is *revolt*, which can cause a critical change in one cycle to cascade up to a vulnerable stage in a larger and slower cycle. The other connection is *remember*, which facilitates renewal by drawing on the potential that has been accumulated and stored in a larger, slower cycle (modified from Gunderson and Holling 2002).

The preceding commentary addresses ecosystem change from a conceptual perspective. The adaptive cycle is another example of an emergent property associated with the ecosystem level of integration. In Chapter 9 we consider landscape change from a mechanistic viewpoint. The panarchy concept provides connection among the different spatial and temporal scales where change is taking place (Holling 2001, Gunderson and Holling 2002).

ECOSYSTEM SERVICES

The concept of ecosystem services grew from an interest in a science-based approach to managing the environment to enhance human welfare. In a general sense, ecosystem services are "the benefits of nature to households, communities, and economies" (Boyd and Banzhaf 2007). The subject of ecosystem services has an ecological and economic underpinning and includes a multi-topic agenda: the definition and elaboration of what constitutes ecosystem services; how changes in ecosystem services have affected human well-being; how ecosystem changes may affect people in future decades; and response options that could be adopted at local, national, or global scales to improve ecosystem management and thereby contribute to human well-being and poverty alleviation (Millennium Ecosystem Assessment [MA], 2005).

The economics component of the concept of ecosystem services is closely tied to interest in systems for environmental accounting and performance assessment. "Services" are often the units that these systems track and measure. Not surprisingly, several views of what constitutes an ecosystem service have been proposed, and unique taxonomic (typological) classifications exist in the literature (e.g., Costanza et al. 1997, Daily 1997, de Groot et al. 2002, National Research Council 2005, MA 2005, Boyd and Banzhaf 2007). To be useful in environmental accounting systems, ecosystem services must be defined by quantity (units) and price. This constraint requires a precise definition of ecosystem services. To address this critical requirement, Boyd and Banzhaf (2007) offer the following definition: *final ecosystem services* are components of nature directly enjoyed, consumed, or used to yield human well-being. Although simple in statement, this definition greatly constrains the elements included in the taxonomy of ecosystem services.

Three features of this emphasis on *final ecosystem services* are important from a welfare accounting perspective. First, final ecosystem services are end products of nature. This issue is captured in the language "directly enjoyed, consumed, or used." The distinction between end products and intermediate products is fundamental. For example, in commercial fishing, target fish populations are the ecosystem service. The processes that created the water quality (e.g., filtering and aeration) necessary for the fish populations to survive, grow, and reproduce are intermediate services. Second, final ecosystem services are components (i.e., things [entities] or characteristics), not functions and processes. Ecosystem functions and processes are the biological, chemical, and physical interactions among ecosystem components that lead to the production of final ecosystem services, i.e., they are intermediate services and their value is in the provision of final ecosystem services (Figure 4.10). For example, in this definition nutrient cycling is an ecological function, not a service. Removing functions and processes greatly reduces the number of taxonomic elements associated with the lists of ecosystem services published in the literature (see references above). Third, there is a fundamental distinction between

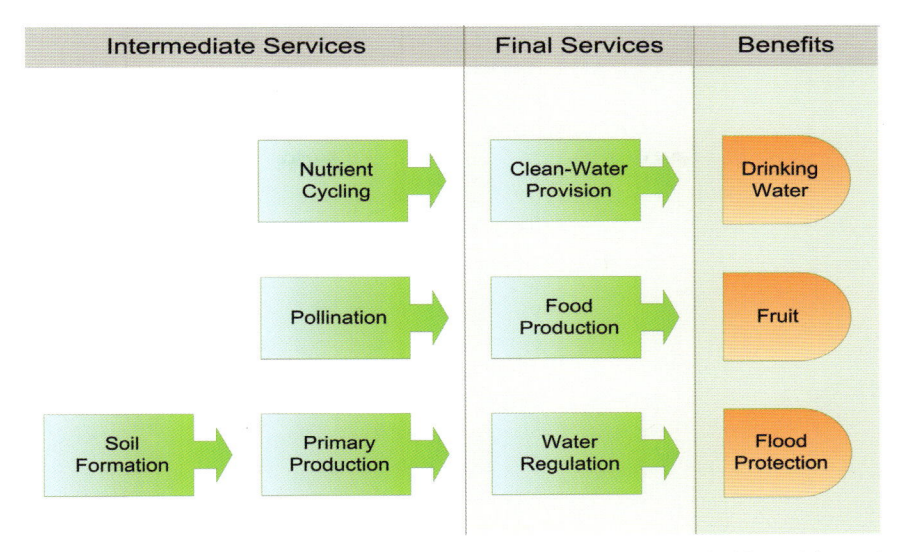

Figure 4.10 — The relation of intermediate and final ecosystem services (modified from Fisher et al. 2008).

the quantity (or physical measure) of ecosystem services and the value of those services. The social value of ecosystem services is spatially explicit. The significance of this feature is that ecosystem services are not spatially fungible or subject to spatial arbitrage. Consequently, the tools and technologies of landscape ecology (e.g., GIS and related spatial analytical procedures) are well suited for evaluating ecosystem services. Table 4.1 provides an inventory of ecological services and benefits derived from them (Boyd and Banzhaf 2007). This inventory is an accounting definition of services.

ECOSYSTEM MANAGEMENT

To *manage* is to take charge of or care of. Herein, we define *ecosystem management* to be the orchestrated modification or manipulation of the basic ecosystem processes (primary production, consumption, decomposition, and abiotic storage) for desired human-defined ends. Generally, the desired ends are the various services (described above) and associated goods.

Ecosystem management is a broad-based concept that blends elements of ecological science, social science, management principals, and technical

Table 4.1 — Inventory of services associated with particular benefits (Boyd and Banzhaf 2007).

Illustrative Benefit		Illustrative Ecosystem Services
Harvests		
	Managed commercial	Pollinator populations, soil quality, shade and shelter, water availability
	Subsistence	Target fish, crop populations
	Unmanaged marine	Target marine populations
	Pharmaceutical	Biodiversity
Amenities and fulfillment		
	Aesthetic	Natural land cover in viewsheds
	Bequest, spiritual, emotional	Wilderness, biodiversity, varied natural land cover
	Existence benefits	Relevant species populations
Damage avoidance		
	Health	Air quality, drinking water quality, land uses or predator populations hostile to disease transmission
	Property	Wetlands, forests, natural land cover
Waste assimilation		
	Avoided disposal cost	Surface and groundwater, open land
Drinking water provision		
	Avoided treatment cost	Aquifer, surface water quality
	Avoided pumping, transport cost	Aquifer availability
Recreation		
	Birding	Relevant species population
	Hiking	Natural land cover, vistas, surface waters
	Angling	Surface water, target population, natural land cover
	Swimming	Surface waters, beaches

knowledge. *Sustainability*, *integrity*, and *health* are prominent elements of the concept of ecosystem management. *Sustainability* is defined as a function of organization, activity, and resilience of an ecosystem. That is to say, a sustainable ecosystem is one that, through a period of time and in the face of management practices, (1) retains the basic elements of its structure, (2) has processes, which define how the system functions, that operate within normal or expected ranges, and (3) can withstand disturbance and return to the normal (nominal) condition. Sustainability also has a temporal component, which is generally defined by a number of human generations, usually five (Forman 1995). This time frame is a practical boundary that attempts to set a realistic planning horizon. *Integrity* is defined simply to mean the state of being whole, entire, or undiminished. The term can refer to elements of structure or the processes associated with ecosystem function or change. *Ecosystem health*, a term taken from medical practice, refers to the conditional states of ecosystems. By definition, an ecosystem is healthy if it is active and maintains its organization and autonomy over time, and is resilient to disturbance (Coulson and Stephen 2006).

Agriculture and forestry represent the two great ongoing experiments in ecosystem management. Manipulation of the ecosystem processes to enhance biomass production is often a principal goal. For example, *primary production* in agricultural fields has been greatly increased by genetic selection, i.e., "plant breeding." The breeding programs have increased the ratio of the edible to the non-edible part of crop plants. However, the efficiency in photosynthesis has not been improved. *Consumption* in agricultural crops by herbivorous pest species, principally insects, has been greatly reduced through the application of pesticides. The result is a substantial increase in net primary production. *Decomposition* in agricultural ecosystems is influenced by different types of tillage/no tillage practices. These practices also reduce export of nutrients by fluvial- or aeolian-mediated erosion, and, along with fertilizer inputs, influence *abiotic storage*. In effect, managing the basic ecosystems process directly influences the quantity and quality of final ecosystem services associated with agriculture and forestry. Management also can be used as a substitute for intermediate ecosystem services, e.g., purification of water, waste treatment.

EPILOGUE TO THE ECOSYSTEM CONCEPT

So what from this review of basic ecosystem concepts is relevant to our discussion of landscape ecology? First, we are representing ecosystems (cf. ecotopes, sites) as the basic building blocks of landscapes. These building blocks are spatially explicit (i.e., defined by geographic coordinates). Ecosystems have structure (biotic and abiotic components) and function (the basic processes of primary production, consumption, decomposition, and abiotic storage integrated through energy flow and nutrient cycling). Ecosystems change in time and space. The process of change was presented as an *adaptive cycle* that represents ecosystem dynamics to consist of four distinct functional phases, referred to as (1) growth or exploitation [r], (2) conservation [K], (3) collapse or release [Ω], and (4) reorganization [α]. From this basic concept of the adaptive cycle, a general theory of change in complex adaptive systems, referred to as *panarchy*, was described. This theory considers ecosystems to consist of a nested set of four-phase interactive adaptive cycles that occur at different spatial and temporal scales. Two anthropocentric applications of the ecosystem concept were examined: ecosystem services and ecosystem management. Ecosystem services are used in environmental accounting systems. Focus was directed to the concept of *final* ecosystem services, i.e., components of nature, directly enjoyed, consumed, or used to yield human well-being. Ecosystem management was defined as the orchestrated modification or manipulation of the basic ecosystem processes for desired human-defined ends. Generally, the desired ends are the various ecosystem services and associated goods. Although emphasis was placed on the structural/functional perspective, it is important to recognize that ecosystem is also an abstract organizing concept and metaphor for holism.

5

The Landscape in Ecology

OVERVIEW

In considering the ways that the agenda of ecology plays out in landscapes, we need a basic understanding of how landscapes are structured, how they function, and how they change in space and time. This fundamental knowledge forms the foundation for the ecological study of landscapes and provides an entry to a science-based approach for managing landscapes. Accordingly, in this chapter we provide an overview of the scope and bounds of landscape structure, function, change, and management. Our discussion of structure considers both the geometry of landscapes and how organisms living in the landscape environment respond to and perceive the constituent elements (ecosystems). The discussion of function considers the flux of energy, materials, and information within landscapes. The discussion of change is based on mechanism. We examine change from two perspectives: landscape-cover change (alteration of the biophysical attributes of the landscape environment), and landscape-use change (human purpose or intent as applied to the landscape environment). The discussion of management is framed in the context of landscape structure, function, and change. Our approach is to introduce these subjects and provide a topical overview (Figure 5.1). In Section II an individual chapter is devoted to landscape structure, function, and change.

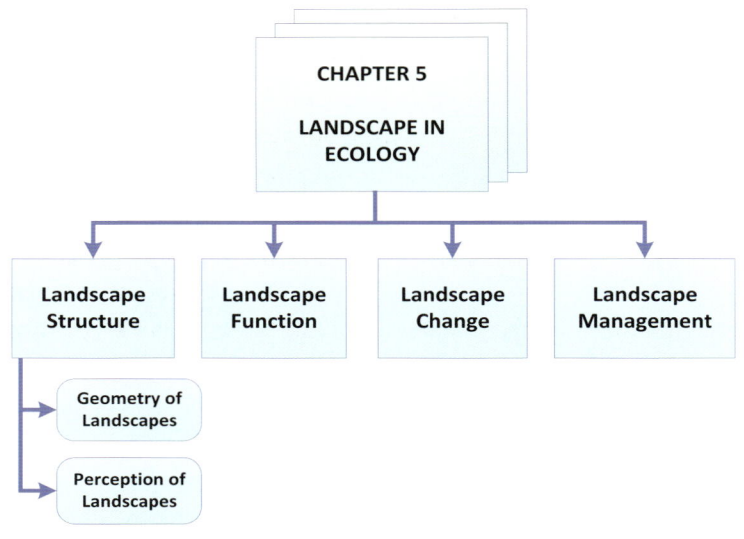

Figure 5.1 — Summary of topics and organization of Chapter 5.

LANDSCAPE STRUCTURE

Landscape structure can be viewed from at least two perspectives. The first perspective emphasizes the *geometry* of landscapes (i.e., the components of the landscape and their linkages and configurations (Forman 1995)) and is an observer-based approach. The second perspective emphasizes *perception* of the landscape (i.e., how an organism perceives and responds to its environment) and is a participant-based approach. Both perspectives are useful in landscape ecology study.

Structure influences the conditions and resources that determine the diversity, distribution, and abundance of living organisms associated with a specific landscape environment. Structure also mediates the flux of energy, materials, and information within this environment, and thereby plays an important role in basic processes associated with landscape function (Wiens 1995). Furthermore, structure influences and is influenced by the mechanisms of landscape change. Landscape ecologists are particularly interested in how living organisms acting as zoogeomorphologists (Butler 1995), ecosystem engineers (Johnston 1995), keystone species, and invasive species change and shape landscape structure.

Landscapes have geometry. When one views a landscape from above, the scene will often consist of an aggregation of different types of ecosystems (elements), e.g., a post oak savanna (Figure 5.2 a and b) or deciduous forest landscape (Figure 2.18) could consist of forest stands, pastures, lakes, roads, etc. To organize the objects forming the landscape so that they can be interpreted from an ecological perspective, landscape ecologists use a simple scheme, referred to as the *Patch/Corridor/Matrix* model. Forman and Godron (1981 and 1986) and Forman (1995) developed, and amplified through explanation, the Patch/Corridor/Matrix model of landscape structure. The seemingly simple, but ingenious, model defines landscapes to consist of three elements: patches, corridors, and a matrix. A *patch* is a surface area differing from its surroundings in nature or appearance (Figure 5.3). This dictionary definition does not carry with it much ecological insight, nor do the common synonyms: piece, scrap, bit. Notwithstanding the difficulty in defining this simple and commonly used word, patches represent a principal structural component of landscapes. Patches have a variety of attributes (size, shape, kind, configuration, number, etc.) and they

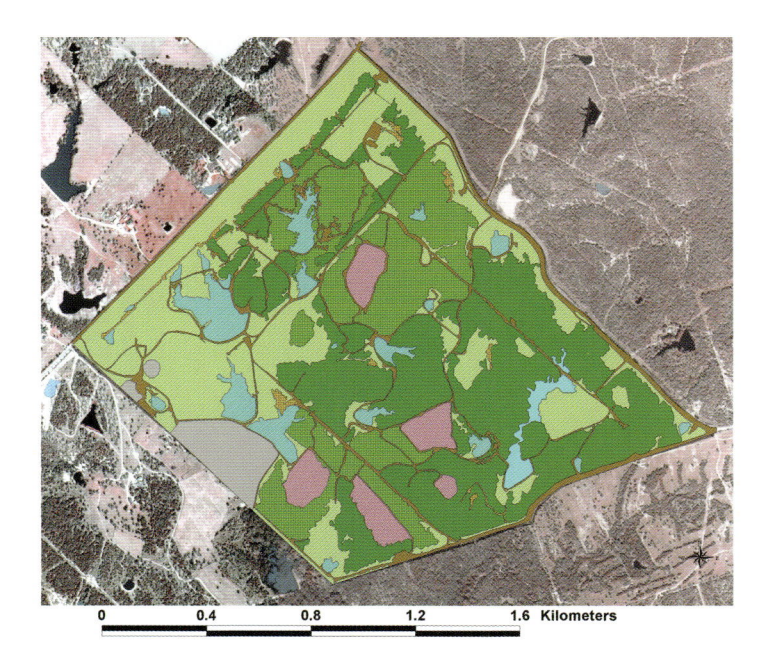

Figure 5.2a — Aerial view illustrating the complex geometry of a post oak savanna landscape occurring in Central Texas (KEL image).

Figure 5.2b — Aerial view of the post oak savanna landscape described in the text where the constituent ecosystems (elements) are labeled. The vocabulary used to identify the individual components is from common parlance and, although descriptive, does not lead to an ecological interpretation of the landscape (KEL image).

Figure 5.3 — An introduced patch, i.e., a surface area differing from its surroundings in nature or appearance (from the "upper lower"- the upper extent of the Lower Peninsula of Michigan) (KEL image).

68 | *The Landscape in Ecology*

originate in several ways (e.g., as a result of a disturbance, such as a fire; as remnants left in managed landscapes; from regeneration of disturbed sites, etc.). The second structural feature of a landscape is the corridor. A *corridor* is a narrow strip of land (or water) that differs from the areas adjacent to it on both sides (Figure 5.4). There are several different kinds of corridors, they originate in a variety of ways, and they have multiple functions in landscapes. The third structural feature of a landscape is the matrix. The *matrix* is the background or cover type of a landscape, i.e., the backdrop in which patches and corridors are imbedded (Figure 5.5). By analogy, the matrix is the dough in a chocolate-chip cookie (the chocolate chips are patches) (Figure 5.6). Generally, the matrix is the most extensive and connected landscape feature. One consequence of landscape structure consisting of multiple components is that the elements (patches, corridors, and matrix) abut (touch or join at the edge or border) one another. The area of abutment creates an additional element of landscape structure, the ecotone. An *ecotone* is a transition area occurring at the interface of two or more distinct landscape components, e.g., different patch types (Figure 5.7). Ecotones are a conspicuous component of landscapes, they are sensitive to environmental change, and they typically are areas where species interactions are intense and biodiversity is high. So technically our basic organizational scheme of landscape structure is the *Patch/Corridor/Matrix/(Ecotone)* (P/C/M/(E) model).

The geometric representation of the landscape environment provides an extremely useful way to organize, classify, observe, and analyze structure. An alternative

Figure 5.4 — An introduced corridor, i.e., a narrow strip of land (or water) that differs from the areas adjacent to it on both sides (Cherolala Skyway, western North Carolina-eastern Tennessee) (KEL image).

The Landscape in Ecology | 69

Figure 5.5 — The matrix is the background or cover type of a landscape, i.e., the backdrop in which patches and corridors are imbedded. Generally, the matrix is the most extensive and connected landscape feature. In this landscape the grassland is the matrix (Lamar Valley, Yellowstone National Park) (KEL image).

approach is to consider how a specific organism perceives and responds to the landscape environment, i.e., the *functional heterogeneity* of the landscape (Kolasa and Rollo 1991). This approach is participant-based and founded on fundamental knowledge of the life cycle, life history, and behavior of the species of interest (see Figures 2.15 through 2.17). This

Figure 5.6 — By analogy, the matrix is the dough in a chocolate-chip cookie (the chocolate chips are patches) (KEL image).

concept is useful in evaluating suitability of a landscape for individual species, it provides a means for evaluating the impact of an organism on the landscape environment, it facilitates investigations of interaction among multiple taxa within the landscape environment, and it serves as a guide for knowledge-based modification and manipulation of landscape structure (i.e., management).

LANDSCAPE FUNCTION

Armed with our basic understanding of and vocabulary for describing landscape structure, we are now in a position to consider landscape function. *Landscape function* deals with the flux of energy, materials, and information within and among the component ecosystems (elements) forming the landscape. As an integrative science, ecology seeks ways to organize for generality. The fundamental question that underpins the

Figure 5.7 — An ecotone is a transition area occurring at the interface of two or more distinct landscape components, e.g., different patch types. In this figure the ecotone occurs at the interface of grassland and forest patches (KEL drawing).

---

subject of landscape function was posited by Reiners and Driese (2004) as follows: "How do the events or conditions in one area of an environmental domain lead to the transmission of effects across that domain to produce an ecological consequence somewhere else?" The answers to this question are the substance of their text and our statement on landscape function.

The general concept of propagation of ecological effects across environmental space is illustrated in Figure 8.2. There are four basic components to this model of transport: an initiating cause, an entity, a propagation vector, and a consequence. The *initiating cause* is defined as the factor that brings about an effect or a result. Examples of initiating causes include atmospheric disturbances (thunderstorms, hurricanes, tornados), fire, herbivory, pathogen infection, etc. A cause can be characterized by a number of different descriptors: e.g., it can be biotic or abiotic in origin; it can be distributed, targeted, diffuse, or patchy in space; it can be frequent, rare, or periodic in occurrence; etc. The initiating cause leads to the movement of some entity (dust, sound, gases, spores, etc.) from its origin to a point of deposition in another place in the landscape. The *entities* (i.e., what is transported) can be classed generally as matter, information, and energy. Specific entities occur in a variety of forms, e.g., photochemical oxidants, pollen grains, pathogen spores, insects, lightning, etc. The *propagation vector* is the means for transport or conveyance of the entity. Suitable synonyms are *carrier* or *conveyor*. There are several different types of propagation vectors, e.g., fluvial (water), colluvial (mass movement), and aeolian (wind) transport; tidal currents; animal movement; etc. The effect of vector-mediated movement of an entity from one place to another within the landscape is referred to as a *consequence* (the right arrow in Figure 8.2). Furthermore, the consequence may become another initiating cause that leads to secondary or tertiary effects (Reiners and Driese 2004). In some instances the consequence of transporting an entity can have a significant impact on landscape structure. Landscape function is discussed in detail in Chapter 8.

LANDSCAPE CHANGE

The topic of landscape change follows directly from our discussion of landscape function. The two subjects are inseparably interrelated. *Landscape change* deals with the alteration of structure (components of the landscape and their linkages and configuration) and function (the flux of energy, materials, and information) of the landscape environment over time. Our emphasis in landscape change centers primarily on the consequences or effects brought about by the operation of the processes responsible for propagation of entities within the landscape environment (Figure 8.2). The various outcomes of the processes that cause change in landscapes are scale dependent. The processes themselves are scale invariant (Sanderson and Harris 2000).

Concern for change in landscapes is often triggered by situations where the rate is faster than normal, e.g., the 1988 forest fires in Yellowstone National Park (Figure 5.8); the magnitude is greater than expected, e.g., the 1980 eruption of the Mount St. Helens volcano (Figure 5.9); the effects on resource values are beyond tolerable levels, e.g., outbreaks of the southern pine beetle, *D. frontalis* (Figure 5.10); and/or the trajectory of landscape succession is changed, e.g., the impact of balsam woolly adelgid, (*Adelges piceae*) on Fraser fir (*Abies fraseri*) forests in the Appalachian Mountains in the southern United States (Figure 5.11) (Coulson and Stephen 2006).

The knowledge base on landscape change can be organized for discussion in a variety of ways. For our purposes we distinguish between landscape-cover change, which deals with the alteration of the biophysical attributes of the landscape environment; and landscape-use change, which deals with human purpose or intent as applied to the landscape environment. Each of these topics is introduced below and discussed in detail in Chapter 9.

Figure 5.8 — Change in landscapes triggered by natural disturbance is often a concern when the rate is faster than normally expected, e.g., wildfire in the Yellowstone National Park in 1988 (NPS image).

74 | *The Landscape in Ecology*

The basic mechanisms of landscape-cover changes include biogeomorphology, activities of living organisms (other than humans), and disturbances. Biogeomorphology is a concept that emphasizes the two-way interplay between ecological and geomorphological processes. The basic premise of this interaction is that the distribution and abundance of species is often related to the underlying geomorphological landform, while surface morphology may in turn be altered by living organisms. The activities of living organisms (other than humans) can modify the conditional state and resources of the landscape environment. Three functional groups of organisms are particularly

Figure 5.9 — Change in landscapes triggered by natural disturbance is often a concern when the magnitude is greater than normally expected, e.g., 1980 Mount St. Helens volcano eruption (USGS image).

important. The first group includes zoogeomorphologists and ecosystem engineers. Zoogeomorphologists are animals involved in eroding, transporting, and/or depositing or causing deposition of rock, soil, and unconsolidated sediments (Butler 1995). The American bison, *Bison bison* (Figure 5.8), is an example. Ecosystem engineers are organisms that mediate change to the abiotic and biotic environment through their influence on the resources and conditions that define habitats for the community of life that forms the biotic component of the ecosystem. The American beaver, *Castor canadensis*, is a good example (Figure 5.12). The second group includes keystone species. A keystone species is one whose effect on an ecosystem is disproportionately large relative to its low biomass in the community as a whole (Power et al. 1996). Again, the American beaver is an example. An important attribute of a keystone species is that there is no

Figure 5.10 — Change in landscapes triggered by natural disturbance is often a concern when the effects on resource values are beyond tolerable levels, e.g., infestation of the southern pine beetle (Sam Houston National Forest, East Texas) (photograph by R. F. Billings).

Figure 5.11 — Change in landscapes triggered by natural disturbance is often a concern when the trajectory of landscape succession is altered, e.g., the impact of balsam woolly adelged, (*Adelges piceae*) on Fraser fir (*Abies fraseri*) forests in the Appalachian Mountains in the southern United States (photograph by R. F. Billings).

redundancy in the community, i.e., there is not another organism that can be substituted to play the role of the keystone species. The third group includes invasive species. An invasive species is a non-indigenous (non-native) organism that has been introduced into a novel ecosystem, i.e., an ecosystem within a landscape that occurs outside the natural range or potential dispersal distance of the species. Their activities can affect the conditional state or resources associated with the landscape environment. The emerald ash borer, *Agrilus planipennis,* is an example of an introduced invasive species (Figure 5.13).

Figure 5.12 — Ecosystem engineers are organisms that modify, maintain, or create habitat (Jones et al. 1997). The American beaver, *Castor canadensis*, is an example (Independence Pass, Colorado) (KEL image).

In addition to the activities of living organisms, landscape-cover change occurs as a result of natural disturbance. For our purposes a disturbance is a physical force, a process, or an event that produces an effect (consequence) that is greater than average, normal, or expected. The effect of a disturbance is the measurable deviation in the value of an ecological variable in relation to a reference condition. This definition of disturbance requires that there be a reference state on condition, i.e., a normal or nominal state. Examples of atmospheric-related disturbances include hurricanes, tornados, ice storms, etc.

The basic mechanisms of landscape-use change are anthropogenic and for discussion purposes can be classified as activities associated with landscape domestication. Landscape domestication deals with shaping landscapes for human welfare, i.e., directed activities aimed at changing

The Landscape in Ecology | 77

Figure 5.13 — The emerald ash borer, *Agrilus planipennis*, is a non-indigenous invasive species that infests and causes mortality to ash trees (*Fraxinus* spp.). The systematic removal of ash spp. will greatly affect species composition of deciduous forests in North America (photograph by David Cappaert, Michigan State University, Bugwood.org).

the landscape environment to provide for basic human needs – adequate food, water, health, housing, energy, and cultural cohesion. This concept of landscape domestication bundles the myriad of human activities that result in change to the landscape environment (Kareiva et al. 2007). These basic human needs translate directly to landscape-use change in the form of consequential actions directed to agricultural production, natural resource management, construction and destruction of the built environment, systems of commerce, provision of public health, and cultural practices.

The various mechanisms of landscape-cover change and landscape-use change physically alter landscape structure and thereby enhance, modify, or eliminate habitat of living organisms. The consequences are seen as effects on the persistence, distribution, and abundance of living organisms in the landscape environment. The mechanisms of landscape change create the observed patterns of landscape structure.

LANDSCAPE MANAGEMENT

To *manage* is to take charge of or care of. In addition to being a discipline of scientific study, landscape ecology also has a substantial applied component that includes landscape-use management, planning, and design. Collectively, these activities constitute the domain of landscape management. *Landscape management* is the orchestrated modification or manipulation of landscape structure (components of the landscape and their linkages and configurations), function (the flux of energy, matter, and information within and among the component ecosystem), and rate of change (alteration in the structure and function of the ecological mosaic over time). Landscape management is a place-based activity involving discrete human activities enacted on a spatially explicit land area organized as a mosaic of component ecosystems. A specific landscape will have a characteristic ensemble of ecosystem types. The unit of management is the landscape and human modifications and manipulations are enacted or performed on the component ecosystems that make up the mosaic (Figure 5.2b).

So the tasks of the landscape manager, practitioner, or designer in modifying and manipulating landscape structure, function, and rate of change are fundamentally different from the landscape ecologist, who seeks to contribute new knowledge to science. Although managers and scientists have different agendas, both utilize the same landscape ecological knowledge base in the practice of their trades. How to identify and manipulate the processes to create desired patterns is very much a part of landscape ecology research. Implementing this knowledge is the challenge of landscape management.

EPILOGUE TO THE LANDSCAPE IN ECOLOGY

In this chapter we opened the study of ecology in landscapes by examining how they are structured, how they function, and how they change in space and time. The fundamental knowledge base associated with these subjects forms the foundation for the ecological study of landscapes and provides an entry to a science-based approach for management. *Landscape structure* was viewed from two perspectives: *geometry* (the components of the landscape and their linkages and configurations) and *perception* (how an organization perceives and responds to its environment). We introduced the Patch/Corridor/Matrix/(Ecotone) model structure and stressed the usefulness of this organizational scheme for describing and comparing landscapes. The discussion of *landscape function* centered on the flux of energy, materials, and information within and among component ecosystems. We presented a four-component conceptual model that facilitated examination of the propagation of ecological effects across the landscape environment. *Landscape change* was defined as the alteration of structure (components of the landscape and their linkages and configuration) and function (the flux of energy, materials, and information) of the landscape environment over time. Landscape change was addressed from two perspectives: landscape-cover change, which dealt with the alteration of the biophysical attributes of the landscape environment; and landscape-use change, which dealt with human purpose or intent as applied to the landscape environment. We concluded by introducing the activity of landscape management. *Landscape management* was defined as the orchestrated modification or manipulation of landscape structure, function, and rate of change.

6

Landscape in Art, Geography, and Landscape Architecture

OVERVIEW

In Chapter 1 we provided a simple definition of landscape: a spatially explicit geographic area consisting of recognizable and characteristic component ecosystems. In Chapter 4 we addressed the ecosystem component of the definition of landscape. With this background we can now proceed with an elaboration of the landscape concept. Accordingly, we consider how various disciplines, in addition to ecology (art, geography, and landscape architecture), use the term *landscape* (Figure 6.1).

A logical beginning for a discussion of the landscape concept is to examine the variety of ways the word is used in common parlance. Casey (2002) summarizes as follows:

> *Landscape:* Literally, "shape of the land" a word deriving from the Dutch *landschap* that signifies (a) a vista or "cut" (hence the *-scape*) of the perceived world, construed as "country" or "land" or "field" set within a horizon; (b) the circumambience [Gibson 1986] provided by a particular place; (c) by extension seascape, cityscape, etc.; (d) genre of painting that is concerned with the material essence of place or region rather than with the precise topography, and with transplacement rather than with transposition (Casey 2002).

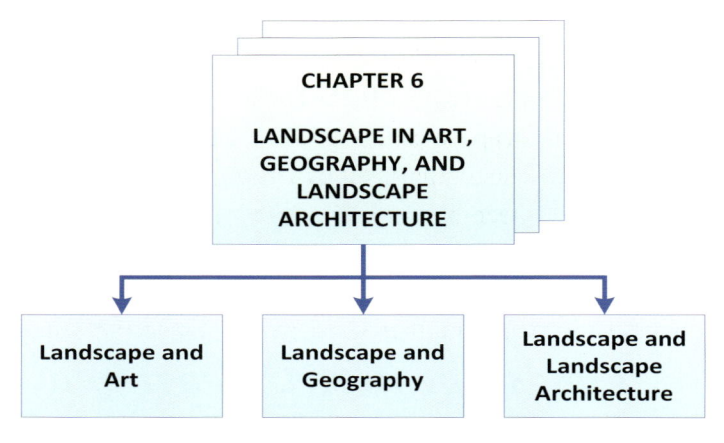

Figure 6.1 — Summary of topics and organization of Chapter 6.

Art, geography, and landscape architecture use the term *landscape* with specialized meaning. These disciplines initially influenced the development of the landscape concept as applied in ecology. Today, there is some degree of reciprocity in the flow of influence among landscape ecology, landscape art, geography, and landscape architecture.

LANDSCAPE AND ART

Most studies in landscape ecology today deal in various ways with images and/or maps. These two media are complementary, and their genesis is in the arts of landscape painting and cartographic representation of place. "Maps (Figure 3.6) give us the measure of place and the relationship between places (quantifiable data) while landscape pictures (Figure 6.2) are evocative, and aim rather to give us the quality of a place or of the viewer's sense of it. One is closer to science, the other to art" (Alpers 1983).

Like landscape ecologists, artists struggle with their description of the term *landscape*. Casey (2002) provides the following comment: "landscape is an instance of what Sartre calls a 'totality detotalized' - it is something that, while experienced as a single whole, is nevertheless not reducible to the sum of its parts (a 'totalization')." In landscape art, land is presented in terms of concrete places. *Places* are the constituent units of every landscape. How places are presented is the substance of landscape art.

A landscape painting both *stands for* and *stands in for* that which it represents. Within the genre of landscape painting there are different *species* (e.g., the landscape of symbols, the landscape of fact, the landscape of fantasy, the ideal landscape, etc. (Clark 1976)) and *modes,* (i.e., the ways that the species are presented - the styles of artists). Because a landscape is envisioned to be an "undecomposable totality" (more than the sum of the parts), there is no criterion of isomorphism for either the species or mode of painting. Furthermore, landscape paintings are *presentations,* i.e., they present in contrast to represent, they show in contrast to replicate (Casey 2002).

In addition to paintings, the other fundamental media for representing landscapes is through maps. Casey (2002) distinguishes between two kinds of mappings. The first type emphasizes accuracy and seeks to characterize with exacting detail the character of specific sites (Figure 3.6). It is a cartographic approach. The second type deals with qualitative characterization of regions and often includes aesthetic criteria. It is a chorographic approach that combines both mapping and painting practices (Figure 6.3).

Figure 6.2 — Thomas Cole, View from Mount Holyoke, Northhampton, Massachusetts, after Thunderstorm (The Oxbow, 1836). The Metropolitan Museum of Art, gift of Mrs. Russell Sage, 1908. This is a faithful photographic reproduction of an original two dimensional work of art.

So what are the commonalities of landscape art and landscape ecology? First, the *places* of landscape paintings are composed of parts of different kinds, i.e., they are characterized by heterogeneity. The heterogeneity in a landscape painting could be described by a landscape ecologist as a cluster of interacting ecosystems. This description would certainly disappoint the artist, and the art scholar as well, but, nevertheless, it would be an accurate characterization of most landscape paintings from a landscape ecological perspective. Second, in landscape painting proper, a natural scene is the primary subject matter (Clark 1976), and it therefore includes both a biotic (living organisms) and abiotic (non-living) environment, i.e., it is composed of the elements of an ecosystem. Third, landscape paintings have spatial scale (range and resolution) (See Chapter 3). The range is usually defined from a point of observation by the field of vision of the artist. The resolution is specified by the artist and represents a desired level of detail. Furthermore, landscape paintings have a temporal component, i.e., we can usually identify the time of day and season of the year the painting presents. The point in chronological history might also be obvious if the painting contains a built environment (e.g., roads, buildings, conveyances, etc.) or human subjects (e.g., period-defined clothing or other "styles").

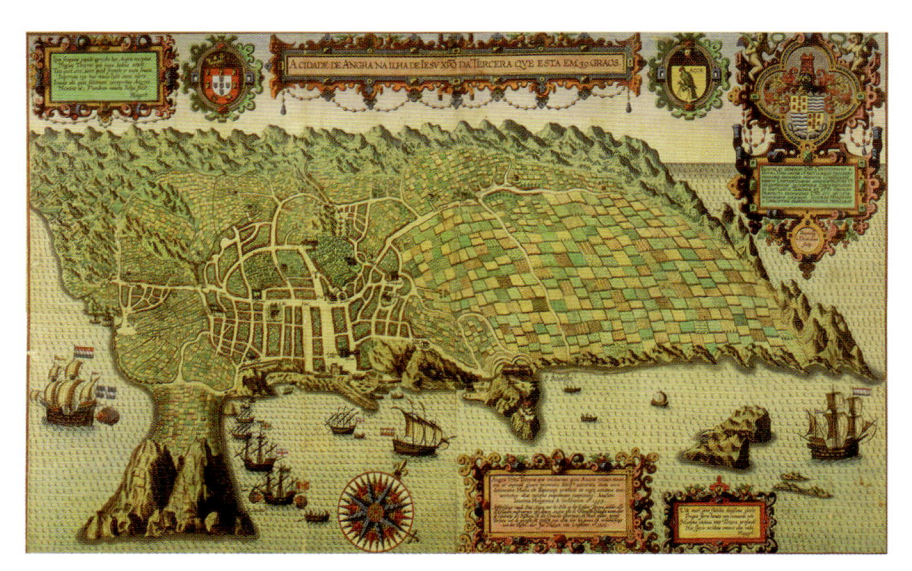

Figure 6.3 — Jan Huygen van Linschoten, View of City of Angra on the Island of Terceira, Azores Archipelago, Portugal, 1595. Maritime Museum, Rotterdam. This is a faithful photographic reproduction of an original two dimensional work of art.

LANDSCAPE AND GEOGRAPHY

The scope and bounds of geography are broadly summarized in Figure 6.4 (Rediscovering Geography Committee, National Research Council 1997), which represents the discipline in three perspectives: place, scale, and domains of synthesis. The domain of *place* is defined as a hierarchy that includes integration in place, interdependencies between places, and interdependencies among places (the horizontal axis in Figure 6.4). *Space* is described using visual, verbal, mathematical, digital, and cognitive approaches (the horizontal axis in Figure 6.4). The *domains of synthesis* include environmental-societal dynamics, which relate human action to the physical environment; environmental dynamics, which link physical and biological systems; and human-societal dynamics, which link economical, social, and political systems (the vertical axis in Figure 6.4) (Rediscovering Geography Committee, National Research Council 1997).

So how does the geographic view of place relate to landscape ecology? Bailey (1996), in his discussion of ecosystem geography, defines the

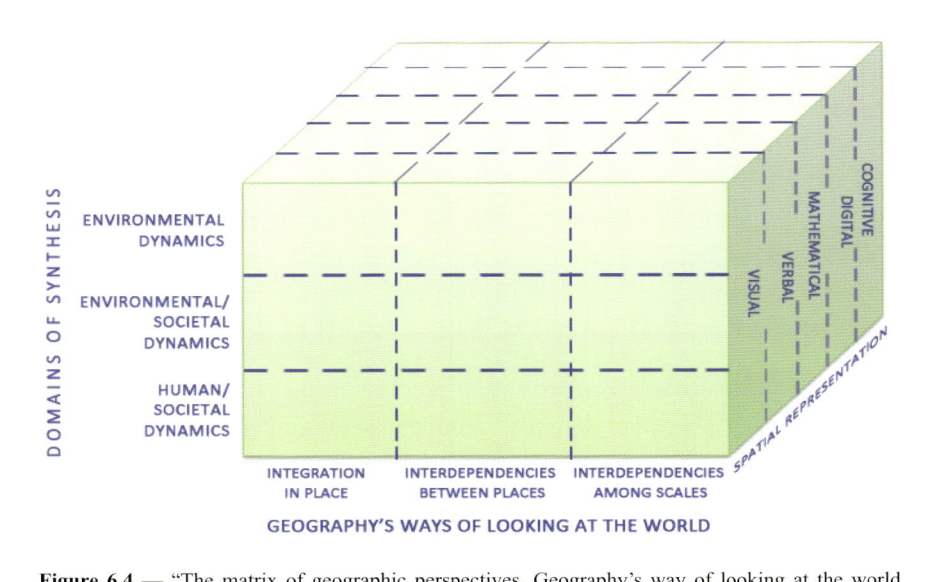

Figure 6.4 — "The matrix of geographic perspectives. Geography's way of looking at the world through its focus on place and scale (horizontal axis) cuts across its three domains of synthesis: human-societal dynamics, environmental dynamics, and environmental-societal dynamics (vertical axis). Spatial representation, the third dimension of the matrix, underpins and sometimes drives research in other branches of geography" (modified from Rediscovering Geography Committee, National Research Council 1997).

domain of *place* explicitly (Figure 6.5) using a hierarchical approach. The site represents the base (sub-) level of the hierarchy. The term *site* is used as a synonym for ecosystem and ecotope, and emphasis is placed on the homogeneous character of this basic unit of organization. The middle (system) level in the hierarchy consists of linked ecosystems referred to as a *landscape mosaic*, i.e., a cluster of interacting ecosystems. This level when viewed from above is frequently a patchwork with a structure defined by different kinds of bounded ecosystems. The top (supra-) level in the hierarchy is the *ecoregion*, which is defined as a geographic group of landscape mosaics. For certain, there are other classifications (and dimensions) used by geographers to describe *place* (see Solon 2005). We have emphasized the system proposed by Bailey (1996) to illustrate how geography views the concept of landscape. Bailey's (1996) model of landscape and place is essentially equivalent to that described by Forman and Godron (1986).

However, there is an additional feature of the geographic view of place that is noteworthy and often ignored or undervalued by landscape ecologists. In addition to the surface geometry, geographers are also concerned with the underlying material or substrate that forms the landscape. This concept of integrating surface geometry and substrate is referred to as *physiography* or *landform* (Figure 6.5). Knowledge of underlying geomorphology of a landscape is of fundamental importance to understanding surface phenomena such as vegetation patterns, distribution and abundance of animal species, land use management practices, etc. (Swanson et al. 1988).

LANDSCAPE AND LANDSCAPE ARCHITECTURE

Landscape architecture is also a multifaceted academic and professional discipline with a broad charge that includes the art and science of analysis, planning, design, management, preservation, and rehabilitation of the land (Ahern 2005). Specific activities associated with the practice of landscape architecture are summarized in Table 6.1. Each of these activities has either a direct or implied landscape ecological underpinning.

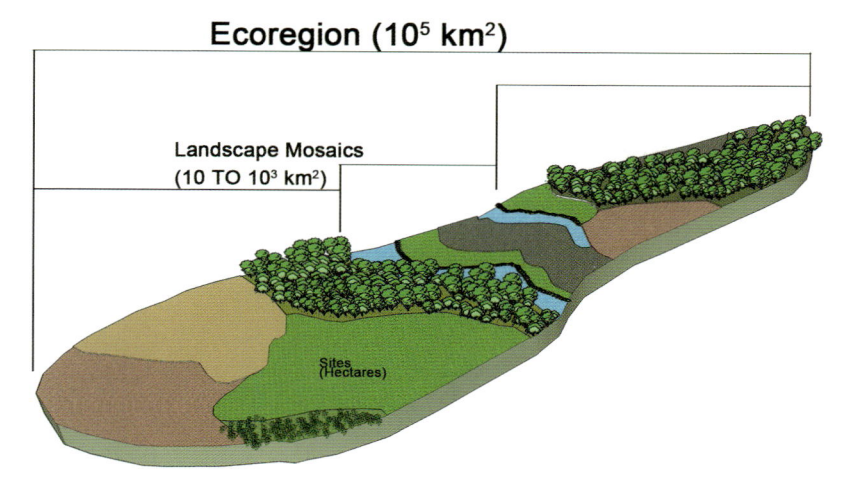

Figure 6.5 — The domain of place from a geography perspective. The site represents the fundamental unit of abstraction. Sites are combined to form a landscape mosaic. Landscape mosaics are aggregated to form an ecoregion (modified from Bailey 1996).

The practice of landscape architecture integrates domain knowledge from other design professions, e.g., *architecture* (which deals with the design of buildings and structures with specific uses such as homes, offices, schools, and factories), *civil engineering* (which applies scientific principles to the design of city infrastructure such as roads, bridges, and public utilities), and *urban planning* (which deals with the broad overview of development for entire cities and regions) (American Society of Landscape Architects <http://www.asla.org>).

The broad-based academic and professional charge of landscape architecture, again, challenges for an inclusive definition of landscape. Marsh (1983) offers the following: *landscape* - "the composite of natural and human features that characterize the surface of the land at the base of the atmosphere. It includes spatial, textural, compositional, and dynamic aspects of the land." This definition accommodates multiple spatial scales, includes both geographical and ecological components, and specifically incorporates human activities. Although stated differently, Marsh's definition incorporates the basic features emphasized in art and geography.

Table 6.1 — The scope and bounds of landscape architecture practice (American Society of Landscape Architects <http://www.asla.org>).

1. ***Landscape Design:*** The concern is with detailed outdoor space design for residential, commercial, industrial, institutional, and public places.
2. ***Site Planning:*** The focus is on the physical design and arrangement of built and natural elements of a land parcel.
3. ***Urban/Town Planning:*** This charge deals with designing and planning cities and towns.
4. ***Regional Landscape Planning:*** An activity that merges landscape architecture with environmental planning, e.g., planning and managing land and water, including natural resource surveys, preparation of environmental impact statements, visual analysis, landscape reclamation, and coastal zone management.
5. ***Park and Recreation Planning:*** An activity that involves creating or redesigning parks and recreational areas in cities, suburban, and rural areas.
6. ***Land Development Planning:*** This activity provides a bridge between policy planning and individual development projects.
7. ***Ecological Planning and Design:*** This activity deals with the interaction between people and the natural environment.
8. ***Historic Preservation and Reclamation of Sites such as Parks, Gardens, Grounds, Waterfronts, and Wetlands.***
9. ***Social and Behavioral Aspects of Landscape Design:*** This activity focuses on the human dimension of design, such as designing for the special needs of the elderly or the disabled.

EPILOGUE TO LANDSCAPE IN ART, GEOGRAPHY, AND LANDSCAPE ARCHITECTURE

Landscape is a common word that belies a simple definition. In this chapter we examined how art, geography, and landscape architecture use the term *landscape* with tailored meaning. Art addresses the two fundamental media of landscape ecology: images and maps. Imagery in art refers to a genre of painting that is concerned with the material essence of place, i.e., the circumambience of a place. Landscape paintings are presentations, i.e., they present in contrast to represent, they show in contrast to replicate. In considering the art of mapping we distinguish between maps that emphasize quantitative accuracy and details of specific sites (cartography) with those that emphasize qualitative and aesthetic features of the landscape (chorography). The geography use of the term *landscape* reflects the broad-based nature of this academic discipline and includes a 3 x 5 x 3 matrix combining perspectives dealing with place, scale, and domains of synthesis. In addition, the geographic concept integrates surface geometry with substrate and includes consideration of *physiography (landform)*, which is often ignored or undervalued in studies of landscape ecology. The *landscape architecture* view of landscape is reflected in the charge of the discipline: the art and science of analysis, planning, design, management, preservation, and rehabilitation of the land. This robust charge results in an encompassing definition of landscape: "the composite of the natural and human features that characterize the surface of the land at the base of the atmosphere."

Our definition of landscape (a spatially explicit geographic area consisting of recognizable and characteristic component ecosystems) emphasizes the geometry of space. The intent of this simple view of landscape is to provide a unit of study and abstraction that could be visualized by fledgling students. The views of landscape from ecology (Chapter 4), art, geography, and landscape architecture are intended to enrich the simple perspective by including the human dimension that leads to a *concept* of landscape.

Preface
Part II — The Substance of Landscape Ecology

In Part I, *FOUNDATIONS*, we examined basic concepts that form the foundation for the discipline of landscape ecology. This discussion was intended to provide a level background for students from the various academic disciplines who typically populate an introductory course in landscape ecology. With this prelude, we can now examine the substance of the discipline in more detail, i.e., at a finer grain. Part II, *THE SUBSTANCE OF LANDSCAPE ECOLOGY*, consists of four chapters (Figure II.1). In Chapter 7, we address *landscape structure*. We begin the discussion by examining the *landscape environment*, a general and non-spatially explicit concept. Next, we consider landscape structure from two perspectives: *geometry* and *perception*. In the discussion of geometry, we begin with an elaboration on the P/C/M/(E) (Patch/Corridor/Matrix/(Ecotone)) model and examine each component in detail. Using this background on the basic components that form a landscape, we next describe how the elements can be clustered into practical functional units that are amenable for studying various topics in landscape ecology, i.e., *ecosystem clusters*. This discussion deals with both configuration and linkages among component ecosystems in the landscape. Often interest in landscapes is in the pattern created by the assemblage of the elements, and we conclude the discussion of landscape geometry by examining the concept of *landscape mosaics*. In addition to providing a template for landscape ecology study, our treatment of geometry is intended to help clarify some of the ambiguity associated with the vague definition of spatial extent of landscapes, i.e., how large or how small is a landscape. In the discussion of landscape perception, we consider the concept of functional heterogeneity and emphasize the utility of this approach for examining how animals (including humans) respond to and perceive the landscape environment. In Chapter 8, we examine the subject of *landscape function*, i.e., the flux of energy, materials, and information within the landscape environment. The discussion is framed within a general model that contains four basic components: *initiating causes*, *entities*, *propagation vectors*, and *consequences*. Each of these elements is examined in detail. We conclude with examples that illustrate the utility of the model. In Chapter 9 we examine *landscape change*, the alteration of structure and function of the landscape environment over time. The goal is to consider the causes and consequences of landscape change. Four topics are considered: landscape-cover change, landscape -use change, how landscape change affects the ecology of living organisms, and the development of pattern in mosaic landscapes. In Chapter 10 we examine *landscape analysis and synthesis*. The purpose of this chapter is to examine how *data, information, and knowledge* of landscape ecology can be used in scientific investigations and for landscape-use management. The goal is to provide a systematic overview of the various approaches used by landscape ecologists for spatial data analysis, synthesis of information, and integration of knowledge. Three topics are considered: analysis of landscapes (pattern and point data), synthesis of spatial data and information (spatial modeling), and knowledge engineering (integrating quantitative data and information with qualitative knowledge).

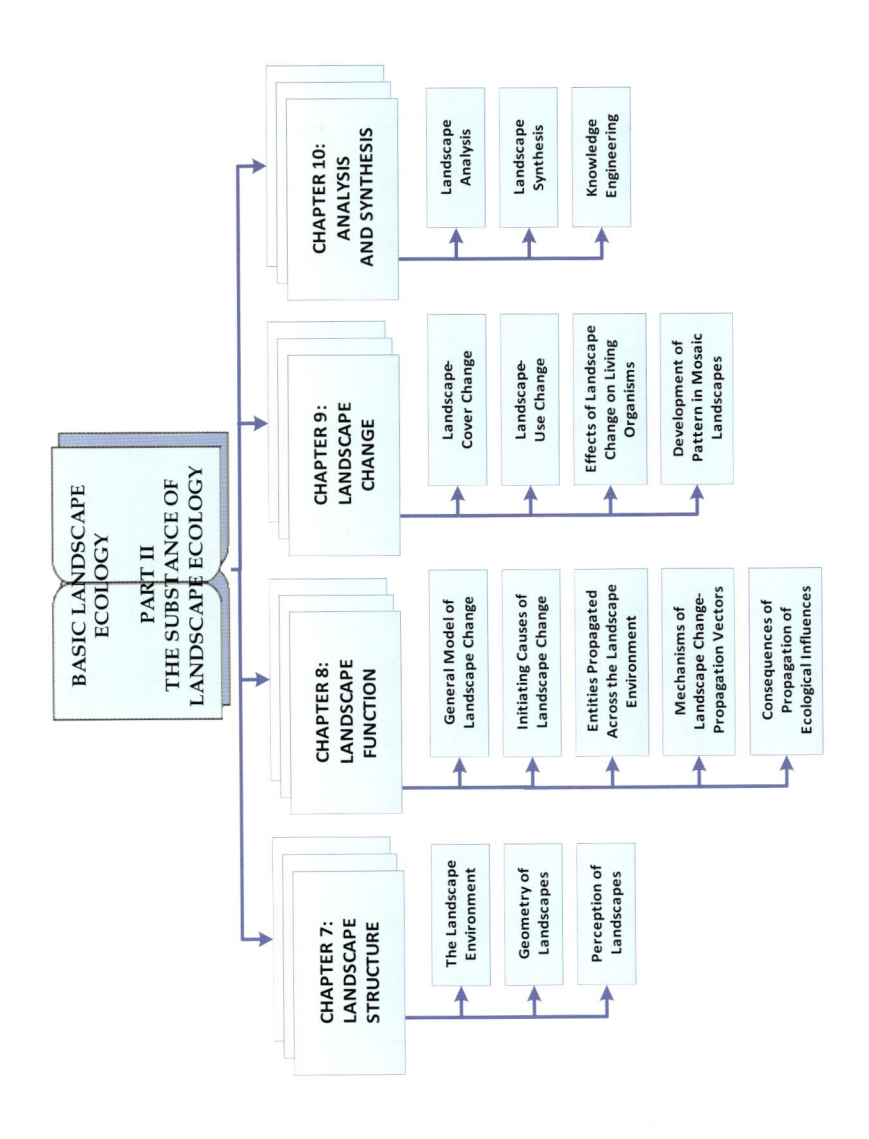

Figure II.1 — Summary of topics and organization of Part II: The Substance of Landscape Ecology.

Preface to Part II | 93

7

Landscape Structure: Environment, Geometry, and Perception

OVERVIEW

Among the significant innovations that accelerated the formulation, evolution, and practice of landscape ecology was the development of a lexicon for describing landscape structure. The foundation for discussion in landscape ecology was provided by Professor R. T. T. Forman in his seminal texts, *Landscape Ecology* (Forman and Godron 1986) and *Land Mosaics* (Forman 1995). In these volumes the vocabulary and concepts of landscape structure were first organized and articulated. These texts present a systematic and broadly accepted vocabulary for describing landscape structure. Herein, our treatment of landscape structure is in large part an abstract derived from these sources.

The goal of this chapter is to examine and illustrate the basic elements of landscape structure. This discussion is divided into three parts: the first deals with a general concept of the landscape environment, the second deals with the geometry of landscapes (which is an observer-based approach), and the third deals with perception of landscapes (which is a participant-based approach). Our treatment of the landscape environment includes consideration of the conditions and resources that affect living organisms; the physical space occupied by living organisms, i.e., the habitat of a species; and the concept of the ecological niche. Geometry of landscapes can be efficiently described using the Patch/Corridor/Matrix/(Ecotone) (P/C/M/(E)) model of structure, which was introduced in Chapter 5. Each component of the model is examined and illustrated. Often, aggregates of the basic components in juxtaposition to one another form a natural ensemble or cluster (*ecosystem cluster*) that serves as a useful unit for structuring ecological investigations. The utility of this unit is considered. We conclude the discussion of geometry with an examination of *mosaic pattern* in landscapes. The point of emphasis centers on the pattern resulting from the assemblage of landscape elements, rather than the individual components *per se*. The final topic in the discussion of landscape structure deals with perception. In contrast to the observer-based view of landscape structure, perception of the landscape environment is participant-based. We consider the questions: How do organisms (including humans) respond to and perceive the landscape environment, and how does landscape structure influence animal movement (Figure 7.1)?

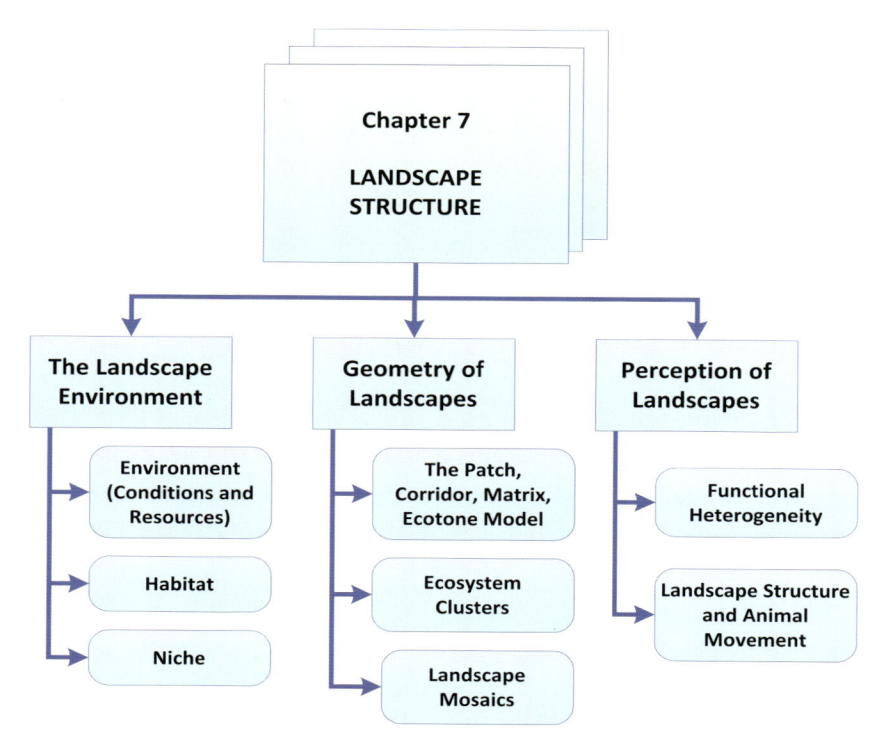

Figure 7.1 — Summary of topics and organization of Chapter 7.

THE LANDSCAPE ENVIRONMENT

At the onset, we defined ecology to be the study of how living organisms interact with their environment. However, we have not discussed the meaning of the term *environment*, which certainly has a variety of specialized interpretations. Nor have we examined the nature of the interaction of living organisms with the abiotic and biotic elements of the environment. Adaptations in morphology, physiology, and behavior facilitate the interaction of living organisms with their environment. The details of the interaction establish where organisms can survive, grow, reproduce, and persist. Herein, we define the term *landscape environment* as follows: all the external conditions and resources (living and nonliving) that affect an organism or other specified system (e.g., a forest landscape) during its lifetime. An environment cannot be described without reference to a particular organism (Lewontin 2000). Environments vary in scale (spatial extent) from global to microscopic. The global perspective is the substance of texts dealing with environmental science (e.g., McKinney et al. 2007). Our focus is on the landscape environment. The following discussion addresses three concepts from basic ecology that are relevant to our examination of the interaction of organisms with the landscape environment, i.e., resources and conditions of the environment, habitat, and the ecological niche.

Conditions and Resources in the Landscape Environment

The conditions and resources in the landscape environment establish the boundaries that define the distribution and abundance of living organisms. The term *condition* is used in a general way to mean simply the state of the environment. Similarly, the term *resource* is defined in the broad sense to be all things consumed by living organisms. In the sections that follow, we expand these definitions.

Conditions

A *condition* is an abiotic environmental factor, which varies in space and time, to which living organisms are differentially responsive (Begon et al. 2006). Conditions define the state of the environment. Examples include temperature, relative humidity, wind speed, pH, salinity, concentration of pollutants, etc. Conditions are not consumed or used up. The importance of environmental conditions to living organisms is defined in terms of their influence on the physiological processes governing survival, growth, and reproduction. For a given environmental factor, e.g., temperature, a species typically will have a broad range of tolerance in which it can survive (defined by high and low thermal death points); a narrower range where it can survive and grow; and an optimal range where it can survive, grow, and reproduce (Figure 7.2).

Several generalizations about the conditional state of the environment and the distribution and abundance of living organisms are important to our study of landscape ecology. First, most living organisms can withstand variation in environmental conditions. Each species has a unique and optimal range of tolerance where the organism will flourish. Second, any condition that approaches or exceeds the limits of tolerance for an organism is said to be a limiting condition or factor. This statement is often referred to as "Liebig's 'law' of the minimum," formulated by Justus von Liebig, a German chemist and the father of the fertilizer industry. Third, lethal conditions have only to occur occasionally to limit the distribution and abundance of a species. In the interim between lethal extremes, species

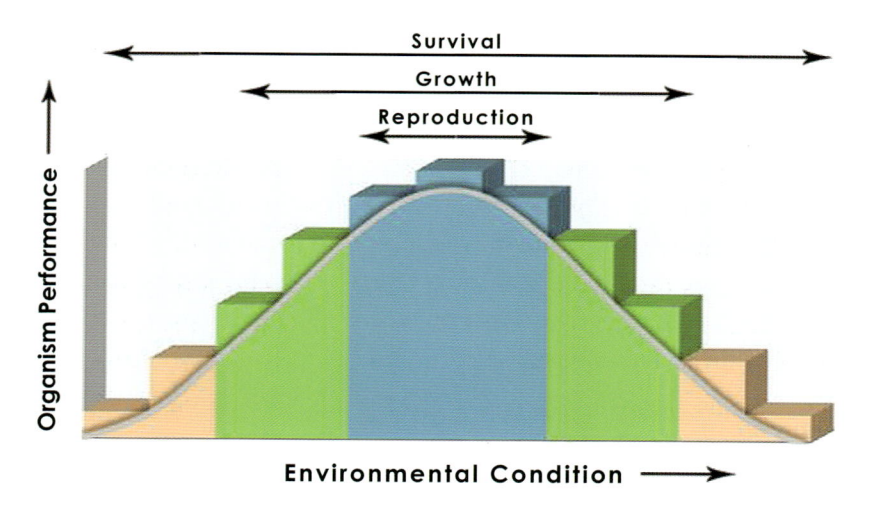

Figure 7.2 — For a given environmental factor, e.g., temperature, a species typically will have a broad range of tolerance in which it can survive (defined by high and low thermal death points), a narrower range where it can survive and grow, and an optimal range where it can survive, grow, and reproduce (KEL drawing).

often expand their range through dispersal or by human introduction. Fourth, the distribution and abundance of a species are limited more often by conditions that are regularly sub-optimal rather than lethal. Sub-optimal conditions often act by altering the outcome of behavioral interactions between different species. The negative effects of sub-optimal conditions are moderated by the morphological adaptations and physiological and behavioral responses of living organisms. Fifth, towards the edge of a species range, it commonly occupies patches where conditions are similar to those found in the center of its range (Begon et al. 2006).

Resources

A *resource* is defined to be all things consumed by an organism. Resources are quantities that can be reduced by the activities of the organism. The word *consumed* encompasses behavior that ranges from an organism sequestering energy to fuel biosynthesis (a green plant photosynthesizing), to ingesting food (e.g., a mosquito taking a blood meal), to selecting a habitat (e.g., feral honey bees occupying a hollow cavity in a tree). Therefore, the resources of living organisms can be broadly bundled into three general classes: energy required for life support, the constituents that make up their bodies, and habitat space associated with their life history.

Energy — In terrestrial landscapes (and aquatic systems), the resource utilization process begins with the sequestration of *radiant energy* from the sun through photosynthesis and the assembly of inorganic ions and molecules by green plants. The energy fixed by photosynthesis and the materials assembled in plant growth are utilized by consumer organisms: herbivores, carnivores, and detritivores. A great deal of the substance of ecology centers on the assembly of inorganic resources by green plants and the reassembly of these packages at each successive stage in a web of consumer interactions (Begon et al. 2006). Generalities about radiant energy as a resource for green plants include the following: first, species differ in their capacity to use radiation as a resource (exemplified in C_3 and C_4 plants); second, radiation is a diurnally and seasonally variable resource; and third, the value of radiation as a resource is critically dependent on and coupled with the supply of water. The energy fixed in primary production is uilized by heterotrophic consumer organisms. In general, there is about a 10 magnitude reduction in energy as it passes from one trophic level to another. For example, if 1 000 kilocalories were consumed by a herbivore, about 100 calories would be converted into herbivore tissue, 10 kilocalories into the first-level carnivore production, and 1 kilocalorie into the second-level carnivore.

Materials — In addition to CO_2 and H_2O associated with photosynthesis, plants require macro (N, P, S, K, Ca, Mg, and Fe); micro (Mn, Zn, Cu, and Bo); and, in some instances, trace (Si, Se, Al, and Co) mineral resources. These resources are extracted independently from the soil by terrestrial plants and from the aqueous medium that surrounds aquatic species. Consumer organisms require many of the same mineral resources, but have the luxury of obtaining essential resources in a bundled form. This circumstance leads to nutritional specialization among different animal species: herbivores, carnivores, and detritivores. Several generalities follow. First, both the quantity (availability) and quality (nutritional value) of food resources are important. Second, many animal species are capable of utilizing a number of different food sources (different plant species, different prey species). Third, many species have a preferred food resource. Fourth, some species are restricted to a single food resource. Phytophagous (plant-eating) animals are often adapted to a particular type of plant module – leaves, flowers, nuts, etc.

Habitat — The term *habitat* is simply defined as the natural environment where a living organism is usually found, i.e., the address of an organism. In the context of resources, habitats are consumed, in that their occupancy quantitatively reduces the number available to other organisms. Habitat is defined by both resources available to a specific organism as well as conditional state of the environment. This simple view of habitat belies the ambiguous use of the concept in the ecological literature. For this reason, we elaborate on the concept in the next section.

Habitat

The concept of the habitat being the place where an organism is usually found is a fundamental organizing principle of ecology. Restated, *habitat* is the physical place where an organism either actually or potentially lives (Kearney 2006). Habitats are described using the biotic and abiotic features of the environment that are thought to be important to an organism. Knowledge of the natural history of a species provides the framework for defining habitat. The places where organisms find food, water, shelter, and space represent good starting points for habitat description. If we know the locations where an organism is found, we can also define the potential range in distribution. In our example of the interaction of the southern pine beetle (*D. frontalis*) and the Red-cockaded woodpecker (*P. borealis*) (Chapter 2), we used habitat targets to relate the two species in physical space, i.e., to illustrate how and why their distributions could overlap. Although this type of characterization of habitat represents a qualitative description that is useful in defining the distribution of the two species, the approach does not provide a means for quantitatively describing the abundance of either (Baum et al. 2005, Mitchell 2005).

Generalities about the concept of habitat include the following. First, a habitat can exist and be described without reference to a specific organism, e.g., a grassland-, a savanna-, a desert habitat. However, ecologists usually have a specific organism in mind when describing a habitat, e.g., American beaver (*C. canadensis*) habitat (Figure 5.13), white-tailed deer habitat (*Odocoileus virginianus*) (Figure 7.5), etc. Second, habitats have spatial locations that are defined by resources and conditions needed by the organism, i.e., it is a distributional concept. Third, there is often a direct relation between the distribution and abundance of suitable habitat sites and the population size and persistence of associated organisms. Fourth, habitat occupancy is variable. In some instances consumption of habitat sites can limit population size, and in other cases there may be many sites that are suitable but unoccupied (Baum et al. 2008). Fifth, how a specific organism experiences a habitat is revealed in the details of its life history.

Niche

The last critical issue in our consideration of how living organisms interact with their environment deals with the concept of ecological niche. The generally accepted view of niche was formulated and articulated more than 50 years ago by Professor G. E. Hutchinson (Hutchinson 1957). The concept of niche is a metaphor (heuristic theory) that addresses how tolerances and requirements of individual species interact to define the conditions and resources needed for survival, growth, reproduction, and persistence, i.e., the concept provides a way of summarizing tolerances and requirements of an organism. Our working definition of *niche* is as follows: a multidimensional space within which the environment (resources and conditions) permits an individual species to survive, grow, reproduce, and maintain a viable population (persist). For an individual species, the dimensions of this space are defined as a function of the inherent tolerances to environmental conditions and resource requirements as well as interactions with other living organisms. Both environmental conditions and resources were discussed in preceding sections. Figure 7.3 illustrates an environment defined by three niche dimensions (temperature, CO_2 concentration, and humidity) for two species, each with different tolerances for these conditions. The ellipsoids represent the *niche breadth*,

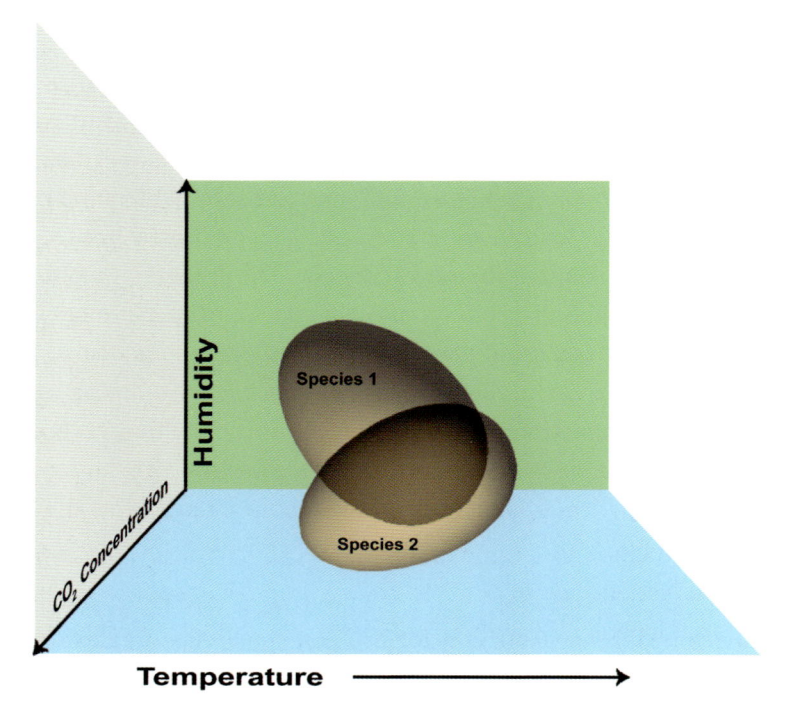

Figure 7.3 — An environment defined by three niche dimensions (temperature, CO_2 concentration, and humidity) for two species, each with different tolerances for these conditions. The ellipsoids represent the *niche breadth*, which is the range of tolerance for these three conditions. The area shared by the two species is referred to as *niche overlap* and in this zone (the shaded area) the species have similar tolerances for the three conditions (KEL drawing).

which is the range of tolerance for these three conditions. The area shared by the two species is referred to as *niche overlap,* and in this zone (the shaded area of Figure 7.3) the species have similar tolerances for the three conditions. We can visualize and illustrate interactions in three dimensions quite easily. However, expansion to include the aggregate of conditions and resources (i.e., the multiple dimensions beyond three) that define the niche for an organism becomes problematic to visualize and impossible to illustrate. In the literature this space is referred to as an *n-dimensional hypervolume*. This term does not hold much meaning for most ecologists, as it is a construct borrowed from a specialized domain in mathematics (see Keet 2006 for a tractable explanation). Nevertheless, the term conveys the idea that the niche of an organism consists of a space defined by multiple interacting conditions and resources unique to each species.

Generalities about the ecological niche, relevant to our study of landscape ecology, follow. First, in a vernacular sense, the concept of the ecological niche deals with how species "fit" into the environment, i.e., how species are constrained by conditions and resources of the environment. Second, a distinction is made between the niche of an organism defined by tolerances to environmental conditions and resources (referred to as *the fundamental niche*) and the niche of an organism influenced by both tolerances and interactions with other organisms (referred to as *the realized niche*) (see Figure 2.4 for the basic types of biotic interactions). Third, for a specific organism, niche dimensions are a subset of the environmental dimensions that directly affect fitness. The term *fitness* is defined as the relative contribution that an individual makes to the gene pool of the next generation. The fittest individuals in a population are those that leave the greatest number of descendants, relative to the number of descendents left by other individuals in a population (Begon et al. 2006, Kearny 2006). The same environment can have different fitness consequences for different organisms, depending on their particular physiology, morphology, and behavior. A specific environmental condition may or may not be a niche dimension for an organism. Unique adaptations of the organism determine which environmental dimensions are relevant (Kearny 2006).

THE GEOMETRY OF LANDSCAPES

From the perspective of geometry, *landscape structure* deals with the components of the landscape and their linkages and configurations (Forman 1995). In this section, we consider three aspects of the geometry of landscapes. First, we examine the P/C/M/(E) model in detail. This model, which was introduced in Chapter 5, provides the basis vocabulary used by landscape ecologists. To be meaningful, the vocabulary must accommodate the extant knowledge base in ecology and to some degree geography. To be useful, the vocabulary has to be universally employed in communication among ecologists and practitioners. The P/C/M/(E) model fulfills these ends. Second, we examine the concept of an ecosystem cluster, which is a unit useful for structuring investigations in landscape ecology. Third, we examine the concept of landscape mosaic. Emphasis is placed on pattern in the landscape, rather than the component entities (ecosystems). Our treatment of the geometry of landscapes has been abstracted from Forman (1995) and this reference provides extensive and detailed explanation.

The Model

When you view a landscape from a distant perspective (as from an airplane or a scenic overlook) what do you see, what vocabulary do you use to describe what you see, and (most important), do you interpret what you see in the landscape in an ecological context, i.e., do you think "ecological thoughts"? You can refresh your memory of the basic components of a landscape (questions one and two) by labeling the elements and describing the landscape illustrated in Figure 4.2. General answers are provided in Figure 5.2a and b. This landscape occurs in a post oak savanna ecoregion (a named derived from the prominent tree species present and a type of climatic zone) and it occurs in Central Texas (Figure 7.4). The third question requires further elaboration. The "ecological thoughts" that might occur to you depend on your specific domain knowledge, interests, or curiosity. For example, if you are an ecologist interested in white-tailed deer (*O. virginianus*) (Figure 7.5) population dynamics, an ecological question might be: What configuration of component ecosystems is required to provide the conditions and resources needed for survival, growth, and

Figure 7.4 — A scenic view of a post oak savanna landscape in Central Texas (US). The ensemble of ecosystems that constitute this landscape are illustrated in Figures 5.2 a and b (KEL image).

Figure 7.5 — White-tailed deer (*Odocoileus virginianus*) are common in post oak savanna landscapes. If you are a wildlife biologist interested in the population dynamics of this species, a landscape ecological question of interest to you might be: What configuration of component ecosystems is required to provide the conditions and resources needed for survival, growth, and reproduction of this animal? (photograph by R.F. Billings).

Landscape Structure: Environment, Geometry, and Perception | 105

reproduction of this species? The white-tailed deer is a common species in a post oak savanna landscape. If you are an entomologist, interested in managing pest insects, an ecological question might be: Which of the component ecosystems are infested with fire ants, and is there variability in population size among the different ecosystems? The red imported fire ant, *Solenopsis invicta*, is an introduced invasive species in this landscape (Figure 7.6). If you are a restoration ecologist, interested in returning the landscape to a pre-settlement condition, an ecological question might be: What plant species should be present, and what patch and corridor configuration would be representative of the original state?

In order to conceptualize and visualize ecology taking place in landscapes, we must have a system for organizing and classifying the constituent components that form the landscape. This classification system will promote the identification and expression of principles and concepts across the agenda of ecology. Recognition of the kinds of landscape components

Figure 7.6 — The red imported fire ant, *Solenopsis invicta*, is an introduced invasive species in post oak savanna landscapes. If you are an entomologist, interested in managing pest insects, an ecological question might be: Which of the component ecosystems are infested with fire ants, and is there variability in population size among the different ecosystems? In this image, entomologists are sampling for the presence of fire ant mounds in an introduced grassland patch (KEL image).

and how they are arranged in context with one another (i.e., as a mosaic of landscape components) opens the gate to the study of landscape ecology. To organize the objects forming the landscape so that they can be interpreted from an ecological perspective, landscape ecologists use a simple scheme, referred to as the *P/C/M/(E)* model. Forman and Godron (1986) and Forman (1995) developed, and amplified through explanation, the P/C/M/(E) model of landscape structure. The seemingly simple but ingenious model defines landscapes to consist of these four basic elements: patches, corridors, matrix, and ecotones. Attributes of these components are described in the following sections.

Patches

A *patch* is a bounded area (i.e., an area that can be defined by coordinates) embedded in the matrix (or another landscape element, i.e., another patch or corridor) that differs from its surroundings in nature or appearance (Figure 5.3). Synonyms include piece, scrap, and bit. Patches are a fundamental structural element in most landscapes and, for certain, their ecological importance is not conveyed by the dictionary definition. The ecological significance of patches in a landscape follows from knowledge of their attributes, origins, and functions in relation to the distribution and abundance of living organisms (Figure 7.7).

Patch Attributes — The simple recognition that patches form a fundamental element of landscape anatomy does not lead directly to an enlightened ecological perspective. However, we can begin to develop an ecological perspective by defining patch attributes and examining how living organisms are influenced by them. The basic patch attributes include the following: size, shape, number, and location. Geographic information systems (GIS), map and image processing software, remote sensing technologies, and spatial metrics (McGarigal and Marks 1995) facilitate the analysis of patch attributes associated with a specific landscape as well as comparison among different landscapes. In Chapter 10 we examine the various metrics used to characterize patch attributes in landscapes. In our case history example dealing with honey bees in pine forests (Chapter 2), an analysis of patch attributes (Table 2.1) was used to explain why the insect was common in what should have been an inhospitable landscape.

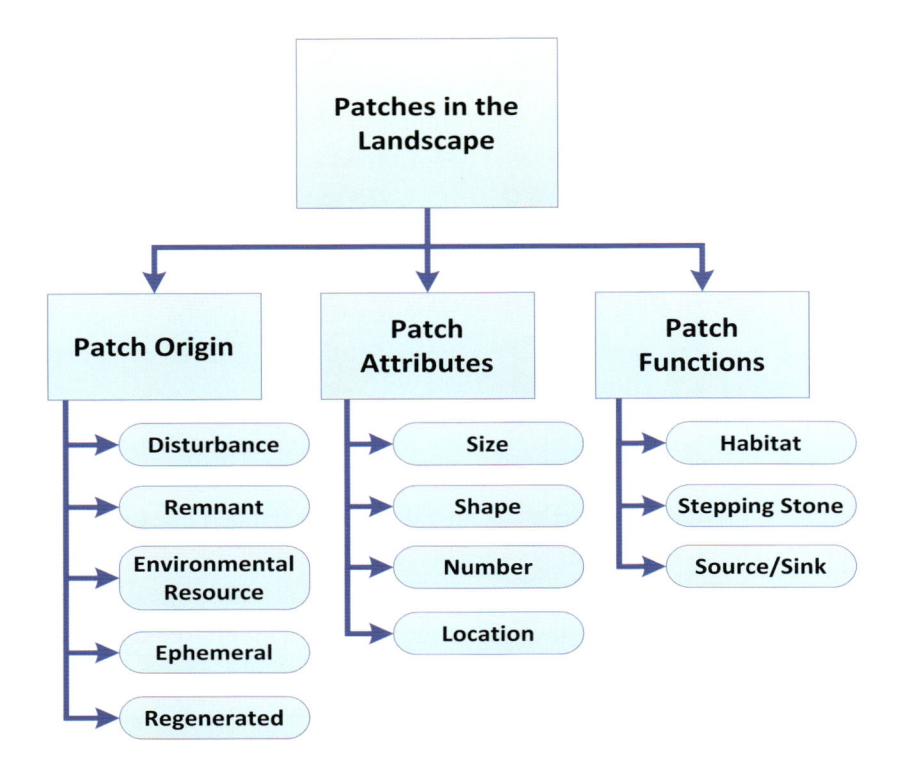

Figure 7.7 — "Mind map" illustrating the basic topics of discussion relating to patches in the landscape.

Once you begin to look for patches in the landscape environment you see that they vary in kind, size, shape, number, and location. The relevance of the specific patch attributes depends on the landscape ecological question you are investigating. However, there are generalizations about patch attributes that are useful in evaluating their contribution to landscape structure. The following section summarizes some of the fundamental generalizations.

Patch size refers to the dimensions of the patch. One of the fundamental questions facing landscape ecologists, planners, and landscape-use managers centers on the importance of patch size, i.e., how large or small do patches need to be. The landscape ecological considerations often focus on details needed to answer questions like the following: Does the patch provide sufficient resources and is the conditional state adequate for a specific organism to survive, grow, reproduce, and persist; how does patch size influence the flux of energy, materials, and information; is the patch size large enough to absorb the impact of disturbance events; etc.? Answering these questions is never a simple matter. For example, in considering how to define optimal patch size in an agricultural landscape, a farmer might be influenced by costs associated with cultivation (tillage, planting, harvesting, etc.) in relation to profits gained at marketing. In modern mechanized agriculture, the concept of economy of scale would favor large patch sizes. However, these units might be much more vulnerable to aeolian erosion, require large inputs of fertilizer and irrigation water, and create a concentrated food source that attracts pest organisms. For non-mechanized subsistence agriculture, energy limitations of humans and their draft animals would dictate small patch sizes.

Patch shape deals specifically with geometry. Common patch shapes include: rounded, convoluted, dendritic, and rectilinear (Figure 7.8 a, b, c, and d). In general natural patches tend to have smooth borders whereas human-created patches associated with forestry, agriculture, and the built environment often have patterns created by straight edges. Human-created patches can be compact or elongated, depending on their intended purpose (Figure 7.8a and d). Uncommon patch shapes (Figure 7.9), typically indicative of human activities, include: spiral, tile, and hourglass designs and shapes. Patch form is often indicative of function. Compact forms are effective in conserving resources, convoluted forms enhance interaction with surroundings, and dendritic (network and labyrinth) forms facilitate transport of materials (Figure 7.8).

Patch number varies among different landscapes. It is an especially significant issue in managed landscapes. Our definition of a landscape (a spatially explicit geographic area consisting of recognizable and characteristic component ecosystems (entities)), implies the presence of multiple patches. In general, patch heterogeneity is a feature of landscapes. However, at a mesoscale and smaller spatial extent, some landscapes are inherently homogeneous in appearance, e.g., tundra, prairie, boreal forests, etc. Furthermore, when we examine photographic images of landscapes, focus is generally directed to obvious patches (see Figure 4.2), and the true character of the landscape may be masked by human uses or the overriding

Figure 7.8 — Common patch shapes include: (a) rounded, (b) convoluted, (c) dendritic, and (d) rectilinear. Compact forms are effective in conserving resources, convoluted forms enhance interaction with surroundings, and dendritic (network and labyrinth) forms facilitate transport of materials (photographs a by Sam Beebe/Wikimedia Commons, b by Maros/Wikimedia Commons, c from MacLean and McKibben (1993), and d from Corner and MacLean (1996), ©2011 Alex S. MacLean/Landslides Aerial Photography).

Landscape Structure: Environment, Geometry, and Perception | 111

vegetation cover (or lack of vegetation). The underlying geomorphology of the landscape may reveal different numbers and patch types than are obvious from a photograph.

Patch location deals with the juxtaposition of patch types relative to one another. The overriding ecological message is that the character and function of the landscape are influenced by content (i.e., the kinds and numbers) and context (i.e., the spatial location) of patches. This reality pervades ecological assessment of landscapes, planning and design of the built environment, and landscape management. Our basic definition represents patches to be inclusions in the matrix. However, patches can be embedded inside other patches (Figure 7.8b), i.e., they can contain an *internal entity*. The forest stand in Figure 7.8b is an introduced patch. Embedded within it is a grassland patch. Furthermore, patches can also be inclusions in corridors. The common occurrence of embedded patches in landscapes requires an adjustment of our basic definition: a *patch* is a

Figure 7.9 — In addition to "art work," uncommon patch shapes, typically indicative of human activities include: spiral, tile, and hourglass designs and shapes. Rock Eagle prehistoric site in Putnam County, Georgia (KEL image).

bounded area, i.e., an area that can be defined by coordinates, embedded in the matrix (or *other landscape element*), that differs from its surroundings in nature or appearance.

Patch Origin — In terrestrial landscapes, patches originate as a function of the operation of both natural and anthropogenic processes. We can distinguish among five basic types of patches: disturbance, remnant, regenerated, introduced, and environmental. The adjectives prefacing the various patch types identify the origin and convey information about their content and character. Following is a brief description of the different types of patches, distinguished by origin.

A *disturbance patch* is a bounded area resulting from a disturbance event that altered the structure and composition of the matrix (or other landscape element). Recall (Chapter 5) that we defined a *disturbance* to be a physical force, process, or an event that causes a sudden deviation in system behavior or a change in system properties. There are as many types of disturbance patches as there are disturbance events that create them. For discussion purposes, the designation *disturbance patch* is often used in reference to a patch resulting from the effects of an exogenous (originating outside the matrix) or endogenous (originating within the matrix) natural process. This restriction is used as a means of distinguishing between disturbance and introduced patches (see definition below). A prominent example is illustrated in Figure 7.10 where the disturbance event creating the patch is a forest fire, likely initiated from a lightning strike. The matrix of the landscape is changed in physical appearance by the addition of the disturbance patch. Biodiversity, edaphic characteristics, microclimate, etc. in the patch will differ from the matrix, following the disturbance event.

A *remnant patch* (adjective, still remaining) is a bounded area left in the matrix (or other landscape element) after a broad scale (i.e., large spatial extent) natural or anthropogenic disturbance event. Figure 7.11 illustrates a remnant patch, a wetland slough ("prairie pothole"), imbedded in an agricultural field. The crop is mustard (Brassicaceae). At the scale of the image, the agricultural field represents the matrix. The agricultural practice is the disturbance event. In this example, the remnant patch remained only because the area it occupied was unsuitable for farming.

Figure 7.10 — A *disturbance patch* in a forest landscape. The disturbance event that created the patch was a forest fire, which was likely initiated from a lightning strike. The matrix of the landscape is changed in physical appearance by the addition of the disturbance patch (USGS photograph).

Figure 7.11 — A *remnant patch*, a wetland slough ("prairie pothole"), imbedded in an agricultural field. The crop is mustard (Brassicaceae). At the scale of the image, the agricultural field represents the matrix. The agricultural practice is the disturbance event. Near Rolla, North Dakota (photograph from MacLean and McKibben (1993), ©2011 Alex S. MacLean/Landslides Aerial Photography).

Figure 7.12 — *Regenerated patch* following selective harvest of a pine stand in the Crossett Experimental Forest (USDA, Forest Service, Southern Research Station) in Crossett, Arkansas. The understory vegetation is a blend of of grasses, pines, and hardwood species (photograph by James Guldin).

A *regenerated patch* (adjective, formed or recreated again) is a bounded area within the matrix (or other landscape element) resulting from the regrowth of vegetation on a previously disturbed site. Figure 7.12 illustrates a regenerated patch in which early successional plants, including seedling pines, have begun to populate a selectively harvested pine stand in a forest landscape. In the ecosystem adaptive cycle scenario (Chapter 4) this patch represents a snapshot in the early exploitation phase.

An *introduced patch* is a bounded area within the matrix (or other landscape element) resulting from landscape domestication (Figure 7.13). Shaping the landscape to accommodate human needs creates a myriad of introduced patch types, each unique in their detail. Introduced patches associated with the built environment, agriculture, and forestry practice dominate this genre.

Figure 7.13 — An *introduced patch* resulting from landscape domestication practices. Golf courses represent a collage of introduced patches, e.g., greens, sand traps, ponds, etc. (photograph by <http:// www.flightimages.co.uk/>).

An *environmental patch* is a bounded area where the resources and conditions differ from the surrounding matrix (or other landscape element) (Figure 7.14). Environmental patches occur as a consequence of a unique conditional state and clumped distribution of environmental resources. Generally, environmental patches are not caused by a disturbance. The living organisms associated with the environmental patch differ from those found in the surrounding matrix because the conditions and resources associated with the patch are unique. In some instances the environmental

Figure 7.14 — An *environmental patch*. This patch type occurs as a consequence of a unique conditional state and clumped distribution of environmental resources. Generally, environmental patches are not caused by a disturbance. Pond in the Hudson River Valley, New York (photograph from MacLean and McKibben (1993), ©2011 Alex S. MacLean/Landslides Aerial Photography).

patch can be ephemeral (Figure 7.15) and the organisms associated with it have specialized adaptations (e.g., short life cycles, tolerances to extreme conditions, etc.) to cope with this situation.

Patch Function — For discussion purposes, we isolated the description of patch attributes (size, shape, number, and location) and patch origin (disturbance, remnant, regenerated, introduced, and environmental). However, in evaluating patches in a landscape context it is important to recognize that they function as an ensemble. The concept of *function* centers on movement of energy, materials, and information in the landscape environment (Chapter 5), and this topic is discussed in detail in Chapter 8. Herein, we are interested in how patches influence and are involved in landscape function. The basic ways that patches influence landscape function is through provision of habitat for living organisms; as stepping stones that network landscape elements; and as sources and sinks for energy, materials, and information.

Figure 7.15 — An *ephemeral environmental patch*. The organisms associated with this patch type have specialized adaptations to cope with the brief temporal duration (e.g., short life cycles, tolerances to extreme conditions, etc.). Pool along Cypress Creek, Texas (photograph by R.F. Billings).

The concept of *habitat*, i.e., the physical place where an organism either actually or potentially lives, was introduced earlier in this chapter. Patch attributes and origin contribute to habitat suitability for specific organisms. For example, when evaluating size, one obvious observation is that large patches may provide a buffered or stable core area (environment) where living organisms are protected from extreme effects of weather-related disturbance events. The organisms that populate such stable habitat sites are often referred to as *K-strategists* (or interior species) and they are characterized by having slow development, low intrinsic rates of increase, long life cycles, large size, etc. In contrast, small habitat patches, not protected from extreme weather conditions, are often populated by species referred to as *r-strategists*. When compared to *K-strategists*, *r-strategists* have fast development, high intrinsic rates of increase, short life cycles, smaller size, etc. These organisms are often referred to as "edge species." The terms *r* and *K* are taken from the logistic equation used to describe population growth. Furthermore, many organisms require more than one type of habitat patch to complete their life cycle or simply to persist. For example, for the Red-cockaded woodpecker, nesting and roosting sites are distinct from the areas where the bird forages for food (see the case history example in Chapter 2 for details). Accessibility to multiple habitat patches is as important as their presence in the landscape environment.

A second important function attributed to patches centers on their role as stepping stones. The term *stepping stone*, in the context of this discussion of function, is a metaphor that addresses how patches connect landscapes. A *stepping stone* is a patch that serves as a means of progress or advancement, usually in reference to movement of living organisms in the landscape. The stepping stone function is particularly important in landscapes where domestication has added heterogeneity through fragmentation, perforation, dissection, shrinkage, and attrition (topics discussed in Chapter 9). Human activities involving these processes can separate populations, isolate critical resource, change habitat conditions, etc. The patches serving as stepping stones may be inadequate for the persistence of a species, but they are suitable refuges that facilitate movement through the landscape.

The third general role of patches in the landscape is as sources and sinks for energy, materials, and information. Again, the concept of sources-sinks, in the context of this discussion of patch function, is a metaphor that addresses concentrations and directional movement. A *source* is a starting point, an area, or a reservoir where output of an entity exceeds input of it. By contrast a *sink* is an end point, an area, or a reservoir where input of the entity exceeds output of it. The entities in the input/output scenario refer specifically to energy, materials, and information. Aeolian or fluvial transport of soil resulting from site preparation and tillage practices applied to an agricultural field (an introduced patch) is an example of materials transport from a source patch (Figure 7.16). Once the crop is planted, capturing solar radiation in this field via photosynthesis represents a patch sink function involving energy (Figure 7.8a). The migration cycle of the monarch butterfly (*Danaus plexippus*) (Figure 7.17) in North America provides an example of both source and sink functions. *Migration* is defined as seasonal movement of a population from one geographical location to another and back again. One route of

Figure 7.16 — Patches can serve as sources of energy, materials, and information. Displacement of soil resulting from site preparation and tillage practices applied to an agricultural field (an introduced patch) is an example of materials transport from a source patch. A haboob (dust storm) rolls in from the Gila River Indian Reservation northward towards the south side of South Mountain in Phoenix, Arizona (photograph by Clayton Esterson).

---

this journey for the monarch butterfly begins in remnant forest patches of Oyamel fir (*Abies religiosa*) in the Transvolcanic Plateau of Central Mexico, i.e., the Mariposa Monarca Biosphere Reserve in the States of Michoacán and México. These forest patches, which provide a moderate climate, become sinks for the immigrating insect during winter months. In the spring these same patches become sources of the emigrating insects that move northward into the eastern United States and Canada. Monarch butterflies are obligatorily dependent on milkweed plants (Asclepiadaceae) to complete their life cycle, and patches of this plant are habitat sinks for the dispersing populations. Several generations (reproductive cycles) of the insect pass before the migrants begin their southward movement back to Mexico, a journey that can range from 1,200 to 2,800 miles. A remarkable facet of the migration cycle is that the individuals that return to the overwintering sites in Mexico are removed by several generations from the parent population. This example bundles the basic entities of transport in a single living organism, i.e., the movement of energy (the

Figure 7.17 — The migration for the monarch butterfly (*Danaus plexippus*) involves both source and sink functions and bundles the basic entities of transport in a single living organism, i.e., the movement of energy (the metabolic processes of life), materials (the substances that make up the body of the insect), and information (the behavior of the insect embedded in the genetic code that defines the species) (photograph by Bastiaan Drees).

metabolic processes of life), materials (the substances that make up the body of the insect), and information (the behavior of the insect embedded in the genetic code that defines the species).

Corridors

The second structural feature of a landscape is the corridor. A *corridor* is a narrow strip of land (or water) that differs from the areas adjacent to it on both sides (Figure 5.4). The adjacent areas can be the matrix, patches, or another type of corridor. Corridors are a fundamental structural element in most landscapes. Nature creates corridors in the form of streams, ridges, and animal trails. Humans create corridors in the form of roads, power lines, ditches, walking trails, and vegetation plantings (Forman 1995). Corridors play contrasting roles in that they both divide and connect landscapes. As with patches, the ecological significance of corridors in a landscape follows from knowledge of their attributes, origins, and functions in relation to the distribution and abundance of living organisms (Figure 7.18). The basic

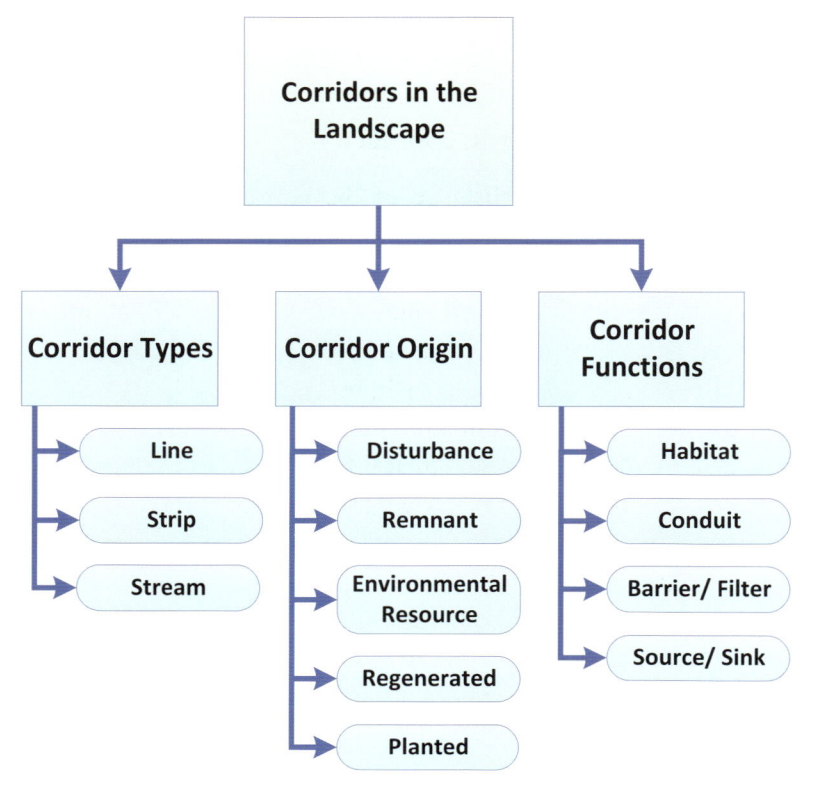

Figure 7.18 — "Mind map" illustrating the basic topics of discussion relating to corridors in the landscapes.

importance of corridors in landscape ecology is reflected in the variety of synonyms that occur in the literature, e.g., landscape linkages, land bridges, wildlife corridors, greenways, shelterbelts, turkey trots, etc. (Hess and Fischer 2001).

Corridor Attributes — We can distinguish among three basic attributes of corridor anatomy: width, composition, and length. These attributes are highly interrelated. The specific values associated with these variables define corridor structure and establish the kinds of functions they can serve.

Width characteristics include a core area, borders with adjacent and bounding landscape elements (i.e., the matrix, patches, and other corridors), and edges (ecotones – see below). The bounding landscape elements are frequently different on each side of the corridor, i.e., the matrix could form one boundary and a different kind of corridor the other boundary. Depending on the dimensions of width, an environmental gradient may occur from side to side across the corridor.

Composition defines the substance of the corridor. In simple natural terrestrial corridors, *composition* follows from the edaphic (soil and surface) characteristics of the landscape environment. In aquatic corridors, water quality, quantity, and flow rates establish the composition of the corridor environment. In constructed corridors, composition is the building material, e.g., cement, asphalt, bricks, etc. In complex terrestrial (Figure 7.19) corridors, with broad width dimension, the central portion may include an *internal entity* (stream, river, road, path, etc.) that adds unique environmental qualities. By contrast aquatic corridors (e.g., rivers and streams), with broad width dimensions, often have lateral structure that also creates a complex and species-rich environment (Figure 7.20). Width and the presence of an internal entity and interior environment are key spatial variables that control terrestrial corridor function. Width and an external or lateral environment are key spatial variables that control aquatic corridor function.

Length characteristics of corridors include variables such as breaks, nodes, narrows, curvilinearity, and connectivity (Figures 7.21 and 7.22). The values these variables take on in a specific landscape greatly influence corridor function. Landscape management practices are often targeted to manipulating these variables to achieve a desired endpoint, e.g., the even flow of traffic in an urban landscape or the protected movement of wildlife in a nature preserve. The length characteristics of corridors are easily identified on aerial photographs and maps, and straightforward procedures for analysis and comparison of them exist (see Chapter 10).

In the preceding discussion, we have represented corridor structure in two dimensions (length*width, m^2). However, some corridors can have volumetric dimensions as well (length*area, m^3) (Figure 7.23). These *tube* corridors are three-dimensional channels of varying size and shape that can occur within the matrix, patches, or other corridors. Flying organisms (mammals, birds, and insects) and arboreal, aquatic, and soil species occupy and utilize three-dimensional corridor environments.

Figure 7.19 — A complex terrestrial corridor with broad dimensions can contain an internal entity (stream, river, road, path, etc.) that adds unique environmental qualities. Width, presence of an internal entity, and an interior environment are key spatial variables that control terrestrial corridor function (photograph by R. F. Billings).

For discussion purposes we can bundle information about corridor attributes into three general types: line, strip, and stream. *Line corridors* (paths, roads, ditches) (Figure 7.24) are the simplest type and are usually characterized by a uniform substrate (composition) from side to side. *Strip corridors* (Figure 7.25) have broader width dimensions than line corridors and this added structure can increase functionality (see below). *Stream corridors* (Figure 7.26) are defined by a water course that can vary in size and complexity as it changes along the full reach of a drainage basin. The classification of corridors as line, strip, and stream is a significant concession to simplification. However, the system is useful in that it provides a means for the initial description of corridor anatomy.

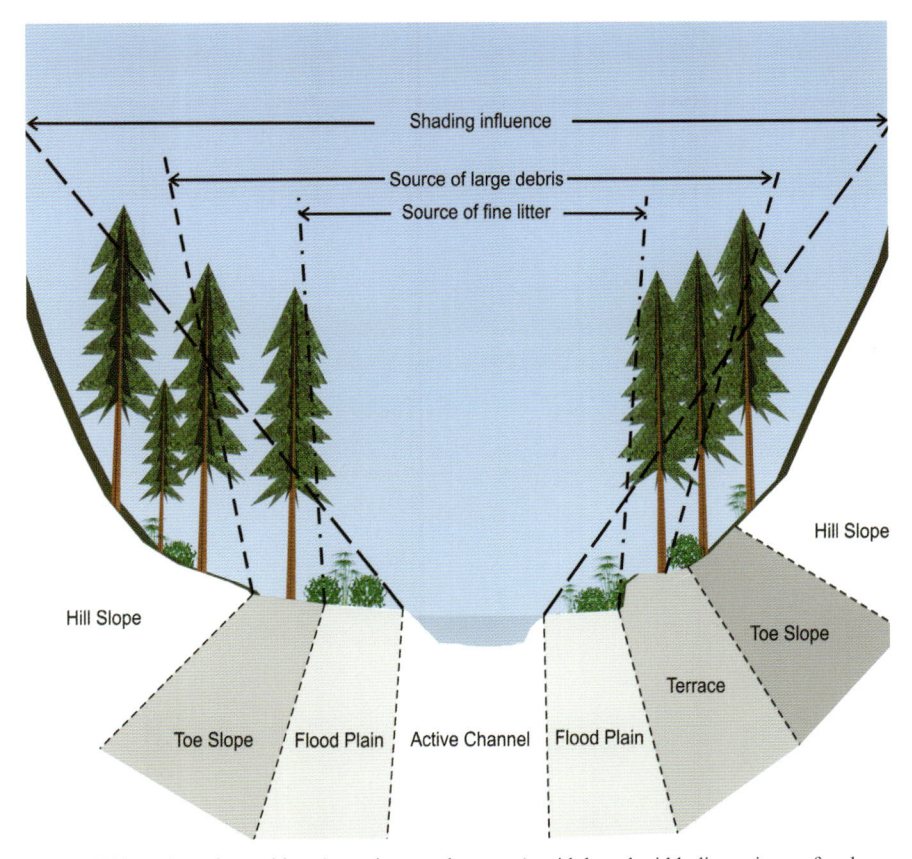

Figure 7.20 — Aquatic corridors (e.g., rivers and streams), with broad width dimensions, often have lateral structure that also creates a complex and species-rich environment. Width and an external or lateral environment are key spatial variables that control aquatic corridor function (modified from Gregory et al. 1991).

Landscape Structure: Environment, Geometry, and Perception

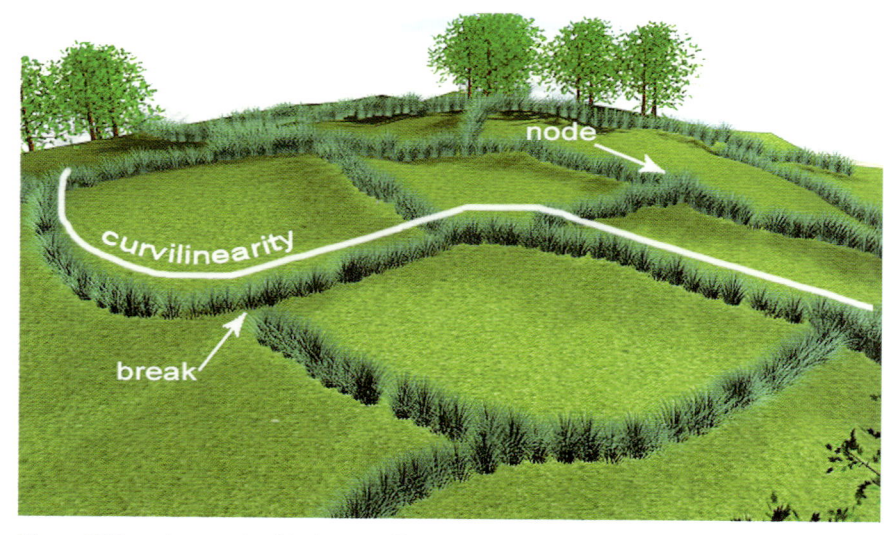

Figure 7.21 — A network of hedgerows illustrates several length characteristics associated with corridors: breaks, nodes, narrows, curvilinearity, and connectivity (KEL drawing).

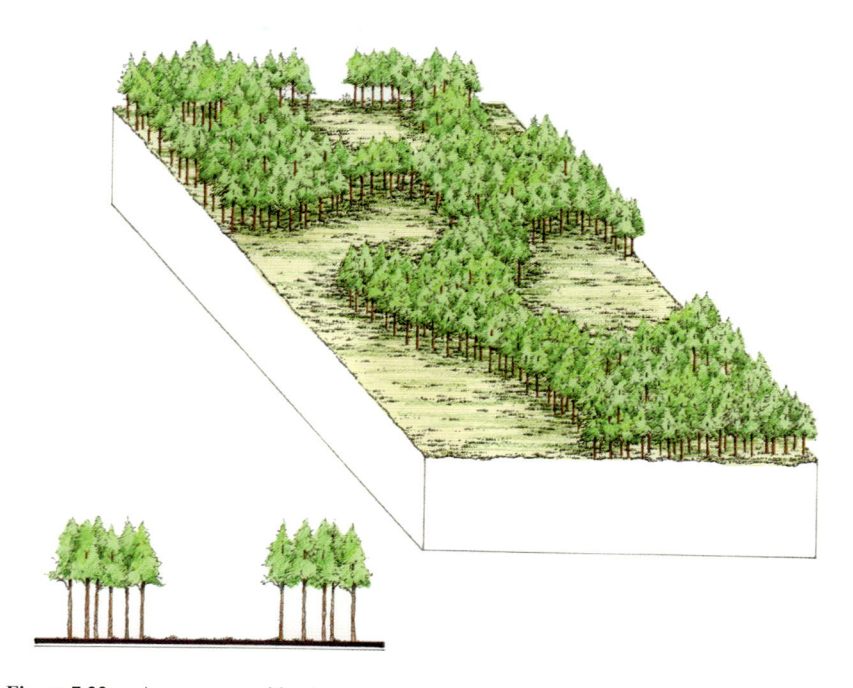

Figure 7.22 — A remnant corridor (see text for definition) illustrating a narrows, where movement of energy, materials, and flow are impeded. An ambush predator would likely select this site to intercept prey (KEL drawing).

126 | *Landscape Structure: Environment, Geometry, and Perception*

Figure 7.23 — Corridors can have three dimensional structure. These tube corridors are channels of varying size and shape that can occur within the matrix, patches, or other corridors. Flying organisms (mammals, birds, and insects) and arboreal, aquatic, and soil species occupy and utilize three-dimensional corridor environments (KEL drawing).

Figure 7.24 — A *line corridor* (path, road, or ditch). This corridor type is usually characterized by a uniform substrate (composition) from side to side (KEL drawing).

Landscape Structure: Environment, Geometry, and Perception | 127

Figure 7.25 — A *strip corridor* has wider width dimensions than line corridors, and this added structure can increase functionality (KEL drawing).

Figure 7.26 — Stream corridor. Tellico River in eastern Tennessee (KEL image).

Corridor Origin — In terrestrial landscapes, as with patches, corridors originate as a function of the operation of both natural and anthropogenic processes. Based on their origin, we can distinguish among five basic types of corridors: disturbance, remnant, regenerated, introduced (planted/constructed), and environmental. Each corridor type represents a narrow strip of land (or water) that differs from the areas adjacent to it on both sides. The adjectives prefacing the various corridor types identify the origin and convey information about their content and character.

A *disturbance corridor* is a bounded linear or curvilinear area resulting from a natural exogenous or endogenous disturbance process (i.e., a sudden deviation in system behavior or a change in system properties). Figure 7.27 illustrates a disturbance corridor created by erosion, i.e., fluvial transport of soil.

A *remnant corridor* is a bounded linear or curvilinear area consisting of parent substrate or vegetation remaining after either a natural or anthropogenic disturbance event associated with one or both of the bordering landscape elements. In Figure 7.28 the remnant corridor is the strip of natural forest vegetation flanked on one side by a lake shore line and on the other by an anthropogenic disturbance resulting from a harvest operation.

A *regenerated corridor* is a linear or curvilinear area composed of natural vegetation that has reestablished following a disturbance event. In Figure 7.29 the regenerated corridor resulted from the residual seed back of pines (*Pinus* spp.) left (or introduced) into the trace left by an outbreak of the southern pine beetle (*D. frontalis*).

An *introduced corridor* is a linear or curvilinear area resulting from landscape domestication practices. Figure 7.30 is a constructed introduced corridor (an irrigation canal) and Figure 7.31 is an example of an introduced corridor extricated from the forest matrix to arrest the movement of a southern pine beetle infestation. This control procedure is similar in approach to creating a fire line to impede the movement of a wildfire.

An *environmental corridor* is linear or curvilinear area where the resources and conditions are unique relative to the adjacent landscape elements. Generally, environmental corridors occur naturally in the environment and

Figure 7.27 — A *disturbance corridor* resulting from a natural exogenous or endogenous disturbance process (i.e., a sudden deviation in system behavior or a change in system properties). The disturbance illustrates erosion caused by fluvial transport of soil (NRCS photograph).

Figure 7.28 — A *remnant corridor* consisting of parent substrate or vegetation remaining after either a natural or anthropogenic disturbance event. In this image the remnant corridor is the strip of natural forest vegetation flanked on one side by a lake shore line and on the other by an anthropogenic disturbance resulting from a harvest operation. Near Lincoln, Maine (photograph from MacLean and McKibben (1993), ©2011 Alex S. MacLean/Landslides Aerial Photography).

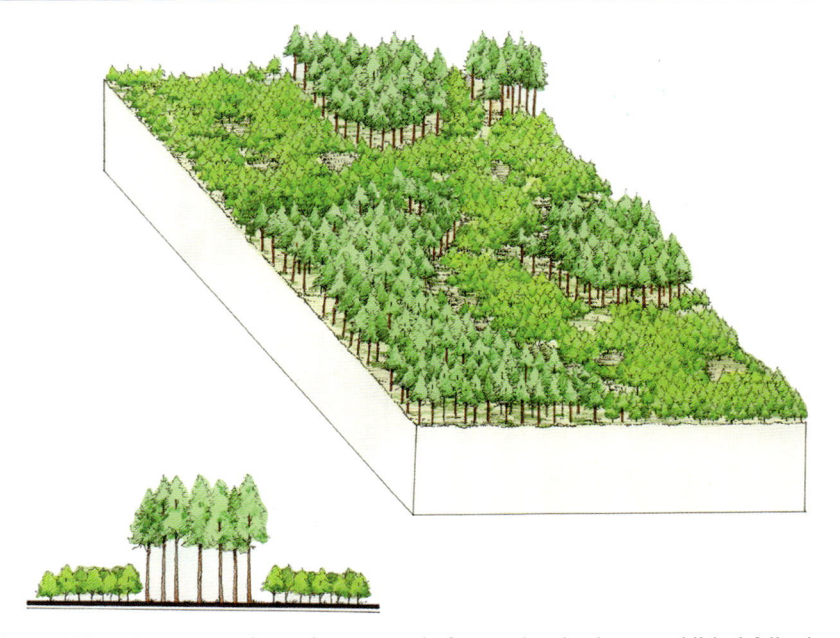

Figure 7.29 — A *regenerated corridor* composed of vegetation that has reestablished following a disturbance event. The trace of the corridor represents the path of herbivory by the southern pine beetle (KEL drawing).

Landscape Structure: Environment, Geometry, and Perception | 131

Figure 7.30 — An *introduced corridor* resulting from landscape domestication: an irrigation canal (photograph from Corner and MacLean (1996), ©2011 Alex S. MacLean/Landslides Aerial Photography).

Figure 7.31 — An introduced corridor resulting from landscape domestication. The corridor was extricated from the forest matrix to arrest the movement of a southern pine beetle infestation. This control procedure is similar in approach to creating a fire line to impede the movement of a wildfire (photograph by R. F. Billings).

are not the result of either disturbance events or landscape domestication practices, and they have distinct fauna and flora associated with them (Figure 7.32).

Corridor Function — Recall that the concept of *function* centers on movement of energy, materials, and information in the landscape environment (Chapters 5 and 8). Herein, we are interested in how corridors influence and are involved in landscape function. The basic ways that corridors influence landscape function is through provision of habitat for living organisms; as conduits for movement; as barriers and filters to movement; and as sources and sinks for energy, materials, and information.

Figure 7.32 — An *environmental corridor* created by a unique assemblage of resources and conditions relative to the adjacent landscape elements (photograph by Lynn Betts/NRCS).

Corridors provide a physical place where organisms either actually or potentially live, i.e., they can serve as *habitats*. Habitat suitability for living organisms is related to the specific attributes of anatomy (composition, width, and length) and the origin or type (disturbance, remnant, regenerated, introduced, and environmental) of the corridor. However, corridor use and activities associated with the adjacent landscape elements also influence habitat suitability. Therefore, the question: "What lives in a corridor?" has multiple answers. In general, species associated with line corridors have to be able to tolerate frequent disturbances and extremes in environmental conditions (Figure 7.33). Strip corridors, which have greater structural complexity and hence habitat diversity, can accommodate a greater variety of species. The diversity of organisms (i.e., the species present and number of individuals per species) associated with aquatic corridors is often used to evaluate water quality. Characteristic faunal assemblages are associated with specific conditional states of aquatic corridors, and for this reason, measurements of species diversity are frequently used to evaluate pollution effects.

Corridors facilitate movement of energy, materials, and information within the landscape environment, i.e., they serve as *conduits*. Movement

Figure 7.33 — Harvester ants (*Pogonomyrmex* spp.) utilize line corridors for habitat in coastal prairie landscapes of South Texas. This insect clears away all vegetation in a 1.0 to 2.0 m^2 area around the entrance to the colony. This corridor environment is inhospitable for most organisms (KEL image).

of entities in the component ecosystems (elements) of a landscape takes place along and within corridors. Again, corridor anatomy, kind/type, and adjacencies influence the conduit function. The ecological interest in conduit function of corridors is addressed from two different but related perspectives: movement of wildlife species in the landscape environment and movement of humans in the built environment. Therefore, the question: "What uses a corridor?" has multiple answers, as well.

At the micro- or mesoscale, wildlife species move in and through the component landscape elements using corridors as conduits. In addition to connecting patches, corridors reduce the impedances to movement and also provide a protected environment for wildlife species as they travel about the landscape. Movement takes place in short discrete time periods and includes behaviors associated with natal dispersal, daily foraging, resting, play, exploration, searching for a mate, etc. This scale of movement is usually within the home range of the organism(s) of interest (Hess and Fischer 2001). The connection, movement, and protection features of corridors are important design considerations for managed landscapes and game preserves.

The conduit function of corridors for wildlife species can include much larger spatial and temporal scales (extent) than movement within and among landscape elements. Migration corridors (see the example above dealing with monarch butterflies) can include distances that extend across continents and involve multiple generations of a species. Of course, these corridors also provide habitat for the migrating species in addition to connecting patches. Protection, preservation, maintenance, and design (tunnels, underpasses, bridges, etc.) (Figure 7.34) of wildlife corridors are all concerns for conservation ecologists.

The conduit function of corridors in domesticated landscapes is a subject of paramount importance and a topic of focus in road ecology. *Road Ecology* deals with the interaction of organisms and the environment linked to roads and vehicles, i.e., it is the discipline that explores and addresses the relationship between the natural environment and the road system (Forman et al. 2003). Roads are fundamental to the organization of domesticated landscapes. The variety and complexity of roads and road networks ranges from multilane highways systems (with associated adjacencies [medians,

Figure 7.34 — Wildlife corridors connect critical habitats in the agricultural heartland of Iowa (NRCS photograph).

shoulders, drainage ditches, vegetation buffers, etc.]) on the high end to unpaved corridors in rural landscapes on the low end. The conduit function facilitates movement of people, goods, and services (flavors of energy, materials, and information). Roads are both attractors and precursors to change in landscapes (Chapter 9). The development of a transportation infrastructure is an essential feature of commerce, which is an activity that underpins landscape domestication. Subjects of particular interest in road ecology include *road density* (the average total road length per unit area of landscape), *mesh size* (the average area or diameter of polygons enclosed by a road network, i.e., the size of the enclosed parcels created by a road network), and *network form* (the explicit spatial arrangement of roads and intersections, i.e., linkages and nodes) (Forman et al. 2003).

Corridors function to impede and filter movement of energy, materials, and information within the landscape environment. The *barrier/filter* functions separate and differentiate the landscape elements on opposite sides of the corridor. Barriers restrict movement (Figure 7.35). Filters provide selective permeability (Figure 7.36). Corridor barriers occur as a result of natural landforms (the Grand Canyon) and water courses (the Mississippi River) and as a consequence of landscape domestication (a multilane highway). The barrier function is used in reference to a specific entity, e.g., the Great Wall of China (Figure 7.35) is a barrier to movement of humans and most terrestrial animals but does not restrict the flight of birds or wind transport of seeds and pollen. The filter function of corridors denotes variable degrees of permeability, i.e., some entities can pass through the corridor while others are impeded (Figure 7.37). Windbreaks serve to filter entrained entities by reducing wind velocity causing suspended particles (e.g., sand) to "drop out." Hedgerows are used as fences to restrict movement of livestock and act as barriers but do not impede movement of small mammals and birds. In forestry, remnant corridors adjacent to riparian zones, lakes, and streams (referred to as streamside management zones – SMZs) are protected because they serve as filters to non-point source pollution of surface water (Figure 7.38).

Corridors can also function as *sinks* and *sources* for energy, materials, and information. Recall that the concept of sources and sinks deals with starting and ending locations, concentration (of ecological entities), and directional movements. A corridor functioning as a source is a starting point for a concentrated ecological entity and movement is outward. Corridors are sources for ecological entities. Living organisms, utilizing the corridor as habitat, are often the focus of interest for landscape ecologists. For example, along the Gulf Coast of the southern United States, large populations of "lovebugs," *Plecia nearctica*, which are actually flies (Diptera: Bibionidae), are associated with road corridors (Figure 7.39). The cultural practice of mowing the grasses along the margins of the road corridors and adjacent drainage ditches provides conditions and resources that are exploited by the insect. Typically, the grass clippings are left in place, and this decaying plant material serves as ideal habitat for the immature stages of the insect. Consequently, large source populations of the insect frequently occur, and emergence in the fall and spring months

Figure 7.35 — Corridors serve as barriers to movement of energy, materials, and species. The Great Wall of China is a barrier to movement of humans and most terrestrial animals but does not restrict the flight of birds or wind transport of seeds and pollen (photograph by Jane Richardson).

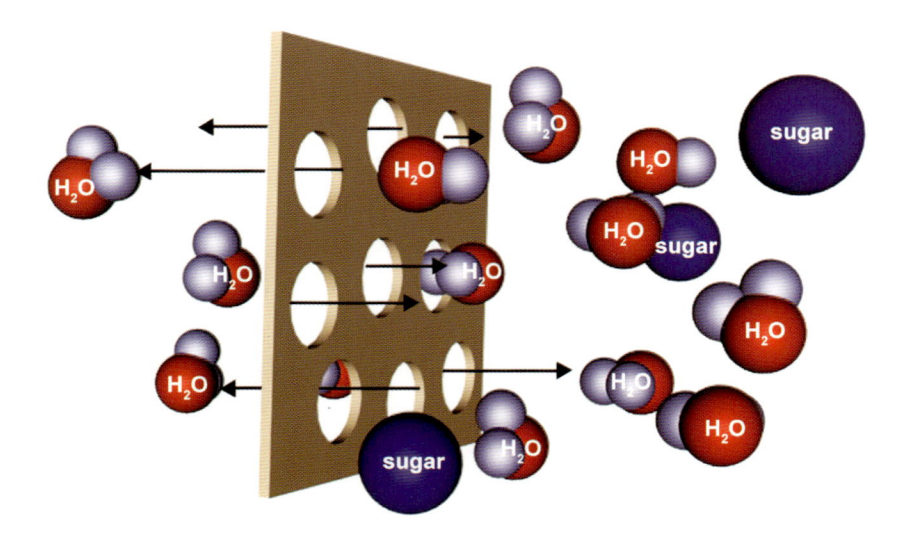

Figure 7.36 — By analogy, the filter function of corridors acts as a semi-permeable membrane: some particles can pass through while others are impeded (KEL drawing).

Figure 7.37 — An extensive farmstead. Windbreak, terraces, grassed waterways, conservation buffers, and conservation tillage combine to protect this farm in Woodbury County in western Iowa. Corridors serve as filters for movement of energy, materials, and information. Windbreaks serve to filter entrained entities by reducing wind velocity causing suspended particles (e.g., dust) to "drop out" (NRCS photograph).

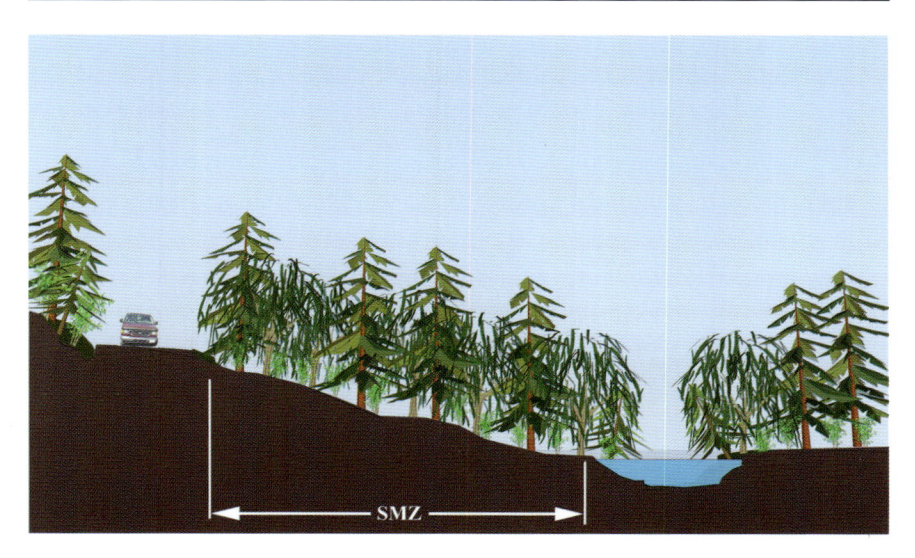

Figure 7.38 — In forestry, remnant corridors adjacent to riparian zones, lakes, and streams (referred to as streamside management zones – SMZs) are protected because they serve as filters to non-point source pollution of surface water (KEL drawing).

Landscape Structure: Environment, Geometry, and Perception | 139

Figure 7.39 — Source populations of "lovebugs", *Plecia nearctica*, which are actually flies (Diptera: Bibionidae), are associated with road corridors. Large populations of the insect frequently occur and emergence in the fall and spring months creates a hazardous and irritating situation for motorists using the corridor as a conduit for movement, i.e., flying insects collide with vehicles, obscuring vision of the driver and damaging the radiator and paint finish (photograph by Bastiaan Drees).

creates a hazardous and irritating situation for motorists using the corridor as a conduit for movement, i.e., flying insects collide with vehicles obscuring vision of the driver and damaging the radiator and paint finish. Road corridors in domesticated landscape serve as a source for a variety of culture-based ecological entities: noise, dust, photochemical oxidants (from vehicles), pollutants, etc. (Forman 1995). The impact of these source entities influences human habitation pattern in urban environments (noise avoidance), selective survival of vegetation (pollution effects created from runoff), wildlife movement behavior, etc. A corridor functioning as a *sink* is an ending point for a concentrated ecological entity, and movement is inward. An obvious example of the sink function of corridors, and an issue of concern in road ecology, is interaction of wildlife species and vehicles, i.e., road kills or faunal mortality resulting from vehicle collision with wildlife species. Behavioral characteristics of wildlife species, in combination with road corridor attributes, interact to produce predictable patterns of mortality (Forman et al. 2003). The sink function of stream corridors was recognized early on by ecologists as the means by which nutrient cycling in ecosystems could be studied (see the case history example in Chapter 2 dealing with disturbance effects on forests). Sink functions of stream corridors include storage and transport of surface water, associated dissolved and particulate nutrients, and living organisms.

Matrix

The third structural feature of a landscape is the *matrix*. The matrix is the background or cover type of a landscape, i.e., the backdrop into which patches and corridors are imbedded (Figures 5.5 and 5.6). The basic attributes used to delineate the matrix in a landscape include area, connectivity, and control over dynamics. *Area* refers to spatial extent, and the matrix is usually the largest unit in the landscape. As such, it serves as the place holder for patches and corridors. *Connectivity* of the matrix refers to spatial continuity, i.e., the degree to which the flow of the landscape is interrupted. A connected matrix is one with minimal interruptions or discontinuities. Connectivity provides cohesion to the geometry of the landscape. *Control over dynamics* refers to the degree to which the matrix influences the flux of energy, materials, and information within the landscape. All of the functional attributes associated with patches and corridors (habitat, conduits, sources/sinks, barriers/filters, and stepping stones) play out at broad spatial scales in the matrix. The idealized generalizations about the matrix, provided by Forman and Godron (1986), include the following: (1) the area of the matrix exceeds the total area of any other landscape element type present, (2) the matrix is more connected than any other type of landscape element present, and (3) the matrix exerts a greater degree of control over landscape dynamics than any other landscape element type present.

The pragmatic description of the matrix component of the landscape (above) is, again, a significant concession to simplification, taken to provide a means for introducing the subject. Recall that in our definition of a landscape (a spatially explicit geographic area consisting of recognizable and characteristic component entities [ecosystems]), spatial extent is a variable. In landscape ecological studies, the area of interest is often specified and delineated for practical purposes and may include a subset of elements occurring at a spatial scale that does not incorporate the matrix. For example, Figure 1.2 is a highly domesticated landscape dominated by introduced patches associated with forestry and agriculture. At the spatial extent of the image, there is no matrix. By contrast, in the natural landscape illustrated in Figure 5.5, the prairie matrix is easily identified and fulfills the basic criteria: size, connectivity, and control over dynamics. The spatial scale of this image is sufficient to capture the matrix as well as the assortment of patches and corridors that characterize this landscape. The matrix is a fundamental component of landscape anatomy, but its relevance in a specific study, investigation, or project varies as a function of the spatial extent specified for the landscape. If the matrix is not evident (Figure 7.40a), expansion of the scale (increasing the range and reducing the resolution) will usually reveal it (Figure 7.40b).

As with patches and corridors, the ecological significance of the matrix in a landscape follows from knowledge of its attributes, origin, and function in relation to the distribution and abundance of living organisms. In comparison to patches and corridors, the matrix has received far less attention as an element of study by landscape ecologists. A great deal of emphasis in landscape ecology has been directed to examination of the habitat function of patches occurring within the matrix and how they are connected by corridors and stepping stones. Initial studies of species movement ignored matrix structure. When models of species movement within landscapes failed to accurately reflect empirical evidence, matrix function was reexamined (Ricketts 2001, Baum et al. 2004, Bender and Fahrig 2005, Donald and Evans 2006). Among the discoveries from these studies, the following are of particular importance. First, different matrix environments provide varying degrees of resistance to movement of species. A *low-resistance matrix* is one that facilitates high rates of interpatch dispersal. A *high-resistance matrix* is one that promotes low rates of interpatch movement. In the literature the term *matrix permeability* is used to convey the continuous nature of variable movement within the landscape. Second, the degree of resistance associated with a specific matrix type is a function of structural attributes and composition of the matrix environment and adaptations of individual organisms, i.e., matrix resistance will vary among different species. Third, attributes of matrix composition and structure influence movement rates and distances traveled

Figure 7.40 — The matrix is a fundamental component of landscape anatomy, but its relevance in a specific study, investigation, or project varies as a function of the spatial extent specified for the landscape. If the matrix is not evident, expansion of the scale (increasing the range and reducing the resolution) will usually reveal it. The matrix is not evident at the spatial extent of Figure 7.40a. The matrix becomes evident in 7.40b, when the range of observation is expanded and resolution decreased (KEL image).

by individual organisms, which, in turn, affect population dynamics and persistence of species. Fourth, matrix restoration is a focus of landscape management and the goal is to improve habitat connectivity by enhancing corridor and stepping stone function.

Ecotones

The basic model of landscape structure consists of patches, corridors, and the matrix. However, a consequence of landscapes consisting of multiple components is adjacency of the elements, i.e., the patches, corridors, and the matrix abut (touch or join at the edge or border) one another. The area of abutment creates an additional element of landscape structure, the ecotone. An *ecotone* (in the context of landscape ecology) is a transition area occurring at the interface of two or more distinct landscape elements, e.g., different patch types (Figures 5.7 and Figure 7.41). A landscape can have a variety of ecotones depending on the number of landscape elements present. So technically, our basic organizational scheme of landscape structure is the Patch/Corridor/Matrix/(Ecotone) model.

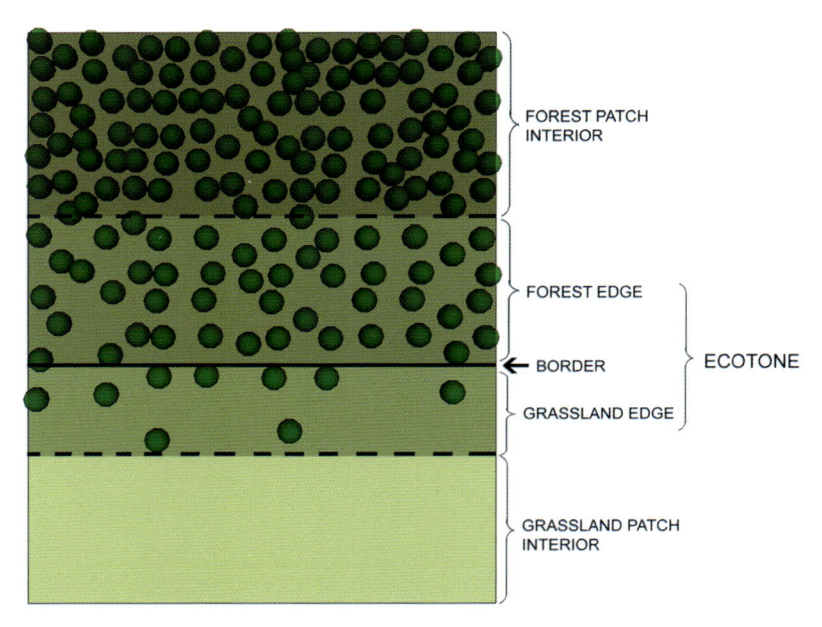

Figure 7.41 — Artist rendition of an ecotone, i.e., a transition area occurring at the interface of two or more distinct landscape elements, e.g., different patch types. This image illustrates the two edge types that form the ecotone (modified from Forman 1995).

The unique structure of each ecotone results from the edge characteristics associated with the adjacent landscape elements, i.e., ecotones do not exist in isolation (Wiens et al. 1985, Holland et al. 1991, Hansen et al. 1992, Forman 1995, Solon 2005). Generally in a landscape ecological study, examination of edge characteristics of adjacent components requires that we adjust the spatial scale of the investigation, i.e., it is necessary to increase the resolution and reduce the range of observation (Figures 5.7 and 7.41) in order to see the specific attributes of the ecotone.

As with the other components of the landscape, ecotones are spatially explicit units of structure. Movement across an ecotone involves a traverse between two different landscape elements, each with unique edge characteristics. The resources and conditions that define the ecotone vary across this traverse in measurable ways. The variation may be small but it can be captured through examination of fine-scale edaphic characteristics. Multivariate methods of analysis can be used to separate and characterize ecotones. This approach also permits comparison of different ecotones (Hargrove and Hoffman 2004).

Ecotones are a conspicuous component of landscapes, they are sensitive to environmental change, and they typically are areas where species interactions are intense and biodiversity is high. A variety of attributes have been identified for ecotones. These attributes result in predictable ecological consequences. Specific attributes of ecotones and their likely ecological consequences are summarized in Table 7.1.

Table 7.1 — Attributes of ecotones and ecological consequence (modified from Despommier et al. 2007).

Biophysical Factors Reportedly Elevated in Ecotones	Ecological or Evolutionary Condition or Change
Species richness and density	Increased frequency of novel species contact
Genetic variability and diversity	Increased opportunity for genetic exchange, genetic novelty
Biological productivity	Increased population growth and turnover rates
Cross and along boundary flows of energy, materials, and organisms	Dispersal and regulation of movement and flow of species, water, and materials
Environmental variability and gradients, habitat heterogeneity	Spatial and temporal environmental variation in biotic and abiotic factors

Landscape domestication creates *anthropogenic ecotones*. Human activities such as forestry, agriculture, and urbanization create unique associations of landscape elements. Examples of anthropogenic ecotones in landscapes include cropland/pasture - natural habitat, forestland-natural habitat, settlement – natural habitat, and a combination of these. The mixture of different kinds of forestland, cropland, pastureland, and natural habitat with different settlement types (urban, suburban, or rural) coalesce into complexes of ecotones at the ecoregion scale (Despommier et al. 2007). Anthropogenic ecotones create new and unique ecological interactions. For example, Despommier et al. (2007) examined their role in the transmission of infectious diseases. Anthropogenic ecotones provided habitat for vectors of disease organisms, e.g., Lyme disease, West Nile virus, rabies, etc. These diseases spill over into human populations and become important in places where exposure had not previously been an issue.

In this discussion our emphasis has been on ecotones as transition areas occurring at the interface of two or more distinct landscape elements. However, the concepts play out at larger spatial scales (extent), e.g., transitions between ecoregions. The multivariate methods for distinguishing between ecotones (Hargrove and Hoffman 2004) were actually developed for evaluating transitions between ecoregions.

Mapping Patches, Corridors, Matrix, and Ecotones

The P/C/M/(E) scheme is pervasive in landscape ecology and the specific vocabulary presented above is used in writing and discussion by both scientists and practitioners. Landscape ecological investigations, landscape planning activities, and landscape management actions frequently involve development of unique maps. Throughout this text, examples of the novel maps, created for specific projects, are presented. However, no standardized mapping symbols are used to distinguish among the different types of elements. Lynch (1996) proposed a landscape ecology symbology, based on a modification of the land-use and land-cover approach advocated by Anderson et al. (1976). The Lynch symbology is presented in Figure 7.42. The system proposed by Lynch (1996) has a basic and unique symbol (referred to as a *parent*) to distinguish patches, corridors, and the matrix.

For patches and corridors there are novel symbols to represent the different types, based on origin, i.e., disturbance, environmental resource, remnant, introduced, and regenerated. These symbols are referred to as *daughters*.

The discipline of landscape ecology has solidified to the point where attention to standardized mapping symbols would be useful

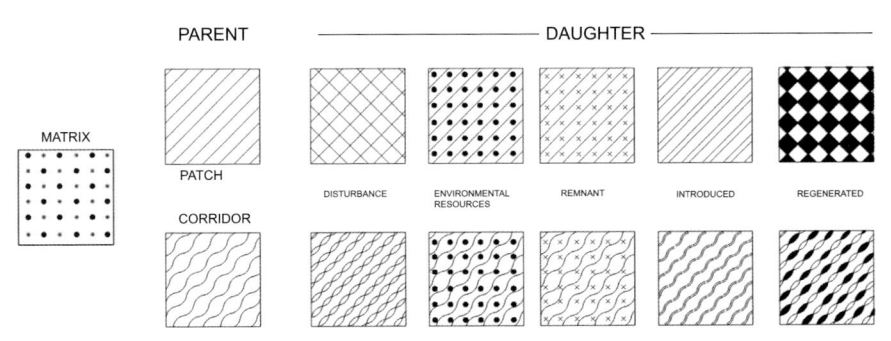

Figure 7.42 — Standardized map symbology to distinguish among patches, corridors, and the matrix (proposed by Lynch 1996).

in communicating among the different disciplines that are using the P/C/M/(E) model for ecological investigations, landscape design, and landscape management. With a basic background in landscape structure, standardized symbols representing the elements of landscape structure (patches, corridors, matrix, and ecotones) convey, in an efficient manner, a great deal of information.

Ecosystem Clusters

Throughout this discussion of structure, we have emphasized that the spatial extent of a landscape is a variable, often specified by the ecologist in response to practical constraints of a specific project, investigation, or study. Many animal species require more than one type of ecosystem for survival, growth, and reproduction. For example with mosquitoes (Diptera: Culicidae), the immature stages (larvae) are aquatic and feed on microorganisms whereas the adults are terrestrial and parasitic, i.e., this insect requires two kinds of ecosystems to complete its life cycle. Ecosystems that are in close spatial proximity tend to be utilized (by a specific organism) most often. For this reason, it is useful to recognize another scheme for organizing landscapes: the ecosystem cluster (Figure 7.43). An *ecosystem cluster* is defined as a group of ecosystems (landscape elements) connected by a significant exchange of energy, materials, and information (Forman 1995). This concept follows from the first law of geography (Professor Waldo Tobler) (Tobler 1970) which states, in essence, that "everything is related to everything else, but near things are more related to each other." For a particular species, dispersal strategies delineate the size of the ecosystem cluster, and details of the natural history define the number needed (Coulson et al. 1999, Coulson and Wunneburger 2000). The terms *geocomplex* (Solon 2005), *meta-ecosystems* (Loreau et al. 2003), and *neighborhood mosaics* (Hersperger 2006) are synonyms for ecosystem cluster.

Landscape Mosaics

The components of landscape structure are aggregated into a mosaic (Figure 7.43). The common use of the word *mosaic* is a picture or decoration made of small pieces of inlaid stone, glass, etc.; something made up of many fragments or diverse elements. The shards (tessera) of glass or stone that make up the mosaic picture are, by analogy, the elements that form a landscape. The *pattern* of elements forming the mosaic picture (and landscape) is the focus of our interest. Analyzing the size, shape, number, and kinds of individual elements that form the image would tell very little about the scene portrayed in a stained glass window. Likewise in investigations of a landscape mosaic, our focus is generally directed to overall pattern, rather than details of the constituent elements.

The landscape mosaic (Hansson et al. 1995) can be envisioned to consist of both horizontal and vertical dimensions. The horizontal dimension, which is referred to as the *chorology (choric)*, deals with the study of the landscape units, i.e., the kinds of landscape elements present that make up the mosaic and their geometric arrangement. The vertical dimension, which is referred to as the *topology (topic)*, deals with the study of landscape attributes, i.e., details of composition — plants, animals, soil, etc. A landscape typically will have a characteristic or specific attribute that is of particular importance and interest, e.g., a pine forest landscape,

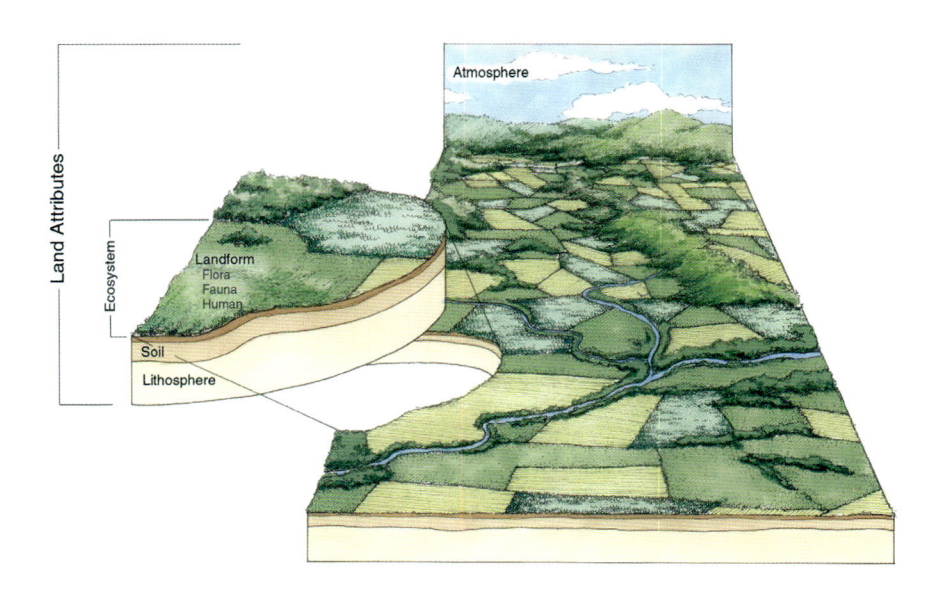

Figure 7.43 — A landscape mosaic consisting of an ensemble of interacting elements (ecosystems). Landscapes have horizontal (chorology) and vertical (topology) dimensions as well as specific attributes. The exploded portion of the figure illustrates an ecosystem cluster, where exchange of energy, materials, and information is accentuated (KEL drawing).

an agricultural landscape, a prairie landscape, or a cultural landscape (Zonneveld 1995, Solon 2005). Each of these features (horizontal dimension, vertical dimension, and characteristic) is illustrated in Figure 7.43. The landscape mosaic occurs as a consequence of the underlying geomorphology, the climatic and edaphic features that influence the living organisms present, and landscape-use practices.

PERCEPTION OF THE LANDSCAPE ENVIRONMENT

In the following discussion of landscape perception we consider two topics. The first deals with the concept of functional heterogeneity, i.e., how living organisms respond to and perceive the landscape environment. The second topic deals with an overiew of the relation of landscape structure to basic patterns of animal movement.

Functional Heterogeneity

The geometric representation of the landscape environment provides an extremely useful way to organize, classify, observe, and analyze structure. Knowledge of the anatomy of a landscape can be put immediately to use in ecological investigations by asking the question: How does an organism (an individual species) understand and use the landscape environment? Landscapes, when viewed from a utilitarian perspective, have *functional heterogeneity* (Kolasa and Rollo 1991) that is determined by the ensemble of elements present and how specific organisms perceive and respond to them. This approach to structuring the landscape environment is participant-based and it is founded on fundamental knowledge of the life cycle, life history, and behavior of the species of interest. The *life cycle* of a species is the sequence of morphological stages and physiological processes that link one generation to the next (Bonner 1993). The components of the life cycle are the same for all members of a species. The *life history* represents the significant features of the life cycle through which a species passes, with particular references to strategies influencing survival, growth, and reproduction. We, as *Homo sapiens*, have the same life cycle, but our individual life histories are unique. *Behavior* is the response of an organism, group, or species to environmental factors. In the context of landscape ecology, our focus is directed to response as a function of perception of the landscape environment. Perception of the landscape environment is mediated through the sensory receptors. For a particular species we are interested in how it uses vision, sound, olfaction, touch, and taste to navigate the landscape environment in search of requisite resources and conditions. Specific details of landscape geometry greatly influence the success or failure of the search.

Knowledge of the life cycle, life history, and behavior of a species can be used to evaluate the suitability of a landscape environment (for a designated species). The ecological knowledge base would need to include information on how the species uses the landscape environment: e.g., how many patch types does it require (food might come from one patch type and shelter from a different one), how large do the patches need to be to provide essential resources, what corridor structure does it use to move within the landscape, how connected do the patches need to be (to avoid exposure to predators or other mortality agents), how permeable is the matrix, etc.? With this type of information, which usually comes from the labors of the autecologist, the landscape ecologist can make knowledge-based inferences regarding the suitability of a landscape environment for a particular species. It might also be possible to identify why a specific landscape would not be suitable for a species and, in some cases, identify the kinds of modifications needed to change or avoid the situation.

We have presented the tenets of functional heterogeneity in the context of an ensemble of elements (ecosystems) that form a landscape utilized by specific organisms. This concept pertains to corridor functionality as well. The conduit function of corridors is frequently challenged by discontinuities along the route of movement. This circumstance, which is particularly important for migration routes, occurs because of inherent breaks and gaps in the corridor or because of landscape domestication practices. Adaptations for flight, water conservation, night movement, etc. can provide *functional connectivity* to corridors.

The knowledge base associated with how an organism understands and uses the landscape environment serves as the foundation for individual-based (object-oriented) modeling of populations. Landscape management is based, in part, on utilization of contextual information. Furthermore, knowledge about how different organisms perceive and respond to the same landscape environment provides a means of studying interactions among different taxa (Coulson et al. 1999) (see the case history of the southern pine beetle and the Red-cockaded woodpecker in Chapter 2).

In summary, the landscape environment can be evaluated in terms of both geometry and perception. The perception-based view leads to the concept of functional heterogeneity, i.e., how a specific organism perceives and responds to the landscape environment. The concept of *functional heterogeneity* requires knowledge of the life cycle, life history, and behavior of species. This concept is useful in evaluating suitability of a landscape for individual species, it provides a means for evaluating the impact of an organism on the landscape environment, it facilitates investigations of interaction among multiple taxa within the landscape environment, and it serves as a guide for knowledge-based modification and manipulation of landscape structure (i.e., management).

Landscape Structure and Animal Movement

A great deal of the scientific and management agenda of landscape ecology deals with the activities and interactions of animals in landscapes. Animal movement is a central theme in this discussion of landscape structure. There are two basic issues important to this point of emphasis: why do animals move and what affect does landscape structure have on movement. Each subject is discussed below.

Why Animals Move

The fundamental reason *why* organisms move in the landscape environment is to secure requisite resources (energy, materials, and habitat) and suitable conditions needed for survival, growth, and reproduction. Herein, we are using the term *movement* to mean a change in place or position. A suitable synonym is *displacement*. Animal movement is a broad-based subject in ecology, and various classification systems exist. For our purposes, we are interested in three types of movement: maintenance movement, dispersal, and migration. Each type of movement is spatially explicit and involves a change in place or position.

Maintenance movement is defined by and is a function of the behavioral motivational states of the animal. Depending on the systematic position of the species (how advanced or primitive it is), various degrees of movement occurs in association with behavior involved in searching for food, water, mates, refuges; parental care; predator avoidance; playing; etc. (Figure 7.44). This type of movement forms the basis of the concept of *home range,* which includes the area where the different behaviors play out for individual species (Roshier and Reid 2003). Maintenance movement is related directly to the concept of functional heterogeneity. In this context, maintenance movement is typically restricted in spatial and temporal extent, i.e., it occurs in small areas and short time frames. If you were interested in observing a particular species of animal and wanted to enhance your chances of seeing it, a good beginning point would be a thorough study of movement patterns associated with the basic motivational states, i.e., where and when does it eat, sleep, play, mate, etc.? With this knowledge you could optimize your observations by going to places where the animal would likely be at a particular point in time.

Figure 7.44 — Animals (in addition to humans) serve as transport vectors. This American bald eagle (*Haliaeetus leucocephalus*) is transporting energy and materials to nestlings (photograph by Ted Coulson, <http://www.tedcoulson.com>).

One outcome of movement is *dispersal*, which is defined to be movement of an individual or a population away from its place of birth or origin. There are several benefits attributed to dispersal behavior, e.g., it facilitates colonization of new habits, exploitation of new food resources, increased likelihood of finding a mate, introduction of heterozygosity into a population, etc. By contrast, dispersal behavior can also result in movement into unsuitable habitats, contact with predators and other natural enemies, exposure to unfavorable weather conditions, reduction in the likelihood of finding a mate, etc.

Migration is movement to and from regular breeding and non-breeding sites. It is usually biannual or seasonal and generally implies breeding site *philopatry*, i.e., the tendency of an individual to return to or stay in its home area, or place of birth (e.g., the example involving monarch butterflies).

Four attributes characterize migratory movement: the breeding and non-breeding sites are spatially separated, the time spent in movement is usually short relative to residency in breeding and non-breeding sites, the journey has outward and return components, and the timing of departure and return is similar each year (Roshier and Reid 2003).

Another useful construct for movement, relevant to landscape ecology, centers on the concepts of home range and geographic range for a species. Essentially, *home range* relates to the discrete area used by an individual or species in its normal activities associated with the behavioral motivational states. *Geographic range* includes the spatial extent of all home ranges of a population or species (Roshier and Reid 2003).

Affect of Landscape Structure on Animal Movement

The surface geometry and the climatic regime associated with a landscape establish in large part what variety of living organisms can be expected to be present. The ecological effects of living organisms in the landscape environment follow from the functional processes that lead to their assembly and movement. Landscape structure influences both assembly and movement in direct and measurable ways. Herein our focus is centers on behavior associated with maintenance movement of animal populations. Following are basic generalizations that summarize how landscape structure guides animal movement (Forman 1995):

1. The landscape mosaic is the template within which animals search for food and habitat.

2. Spatial elements (component ecosystems) of the landscape mosaic vary widely in their suitability for different animal species.

3. Animal species have targets that are suitable patches containing resources and conditions necessary for survival, growth, and reproduction.

4. Animal movement is a function of spatial attributes of each landscape mosaic. Within a specific landscape both the context and content of the spatial elements are important variables influencing movement.

5. The arrangement of suitable and unsuitable component ecosystems forming the landscape mosaic is a major determinant of the movement of animal species.

6. The spread of the animal species is especially dependent on the arrangement of *corridors* that can act as conduits, filters, barriers, habitat, sources, and sinks.

EPILOGUE TO LANDSCAPE STRUCTURE

So what conclusions can be made about the geometry and perception of landscape structure?

1. Our treatment of the subject of landscape structure featured discussion of the landscape environment (conditions and resources in relation to habitat requirements and niche constraints of living organisms); the geometry of landscapes including the basic elements of anatomy (patches, corridors, matrix, and ecotones); and perception of landscapes by specific organisms (functional heterogeneity). The rational for this approach was to link geometry and perception of landscapes through a basic understanding of adaptations of living organisms to the landscape environment.

2. The P/C/M/(E) model (Forman 1995) provides a classification system that facilitates the identification and expression of principles and concepts of landscape ecology across the agenda of ecological science. The model also facilitates discussion of landscape change, function, and management. The model provides a way to think "ecological thoughts" about landscapes.

3. The vocabulary for the P/C/M/(E) model pervades both the scientific literature and popular journalism about landscapes.

4. The ecological significance of the elements of landscape geometry (patches, corridors, matrix, and ecotones) follows from knowledge of their attributes, origins, and functions in relation to the distribution and abundance of living organisms.

5. The P/C/M/(E) configuration for a specific landscape forms the template within which ecological processes operate and subsequent

patterns emerge. From the geometry of landscapes, we can begin to pose important ecological questions, e.g., (1) What processes created the observed pattern; (2) what are the consequences of the pattern on species distribution, diversity, and abundance; (3) how does the configuration of elements influence movement of energy, materials, and information; (4) how do mobile organisms affect landscape heterogeneity, etc.?

6. A standardized symbology for the elements of landscape structure (patches, corridors, matrix, and ecotones) facilitates communication among scientists and practitioners of landscape ecology.

7. The components (patches, corridors, matrix, and ecotones) forming a landscape do not necessarily have to be ecosystems, e.g., an asphalt parking lot (a type of patch) or a paved road corridor in an urban landscape. If it is important to identify whether or not a landscape component is an ecosystem, simply ask the following questions: Is there a biotic community and an abiotic environment and are the processes of ecosystem function (primary production, consumption, decomposition, and abiotic storage) operating, i.e., linked by energy flow and nutrient cycling? Recall that in our definition of *landscape* (a spatially explicit geographic area consisting of recognizable and characteristic component *entities*) there was flexibility to exchange terms like ecosystems, ecotopes, sites, elements, tessera, units, etc.

8. The P/C/M/(E) model accommodates investigations of landscapes at different and appropriate spatial scales. In some instances the useful scale might be a cluster of closely related interacting ecosystems. In other instances, the *pattern* of a landscape mosaic (which becomes apparent only at an expanded range and reduced resolution) might be of interest. The same landscape could be examined at both (and additional) scales. From this discussion it follows that the term *scale* (spatial extent), when applied to a landscape, is a variable. The prefixes micro-, meso-, or mega-scale (Chapter 1, Table 1) are useful in designating and relating general information about the spatial extent of a specific landscape. Nevertheless, the term *landscape scale* is often used in common parlance. The intent is to imply that the land unit or area of interest has a large or broad spatial extent. Although

imprecise, in this context the expression *landscape scale* is no more or less egregious than our use of ecosystem as a metaphor for holism. The general meaning is implied and usually understood.

9. The utility of the P/C/M/(E) model is challenged when the landscape environment is inherently homogeneous, e.g., unbroken tropical or temperate deciduous forests, chaparral, alpine tundra, etc. (Reiners and Driese 2004). Haines-Young (2005) uses the term "fuzzy landscapes" to describe this condition. Furthermore, we can remove heterogeneity in a landscape study by reducing the range and resolution (scale) of observation. It is also noteworthy that normally heterogeneous landscapes can be homogenized through the activities of humans (domestication), e.g., farming and forestry. However, with minor alteration, our basic definition of landscape still applies to the homogeneous conditional state, i.e., a spatially explicit geographic area consisting of a recognizable and characteristic component *entity*.

10. The discussion of landscape geometry (P/C/M/(E) model, ecosystem clusters, and mosaics) and landscape perception has been presented in reference to ecology, but the concepts are directly applicable to issues of landscape architectural design and planning as well.

11. An alternative view of landscape structure centers on perception of the landscape by living organisms. The concept of *functional heterogeneity* of landscapes (how an organism perceives and responds to the landscape environment) is an alternative and useful way to examine landscape structure. Landscapes influence animal movement and therefore directly affect both the distribution and abundance of living organisms in the landscape environment.

12. How structure influences landscape function and the process of landscape change is a fundamental question in landscape ecology and a principal concern for landscape management.

13. Landscape structure influences animal movement. This subject was investigated from two perspectives: why animals move and how landscape structure affects movement. The fundamental reason why organisms move in the landscape environment is to secure requisite resources and suitable conditions needed for survival, growth, and

reproduction. Three modes of movement were examined: maintenance movement, dispersal, and migration. Landscape structure affects both the assembly and movement of living organisms in the landscape environment. General rules, based on the basic elements of landscape structure (patches, corridors, matrix, and ecotones), were examined.

8

Landscape Function

OVERVIEW

Landscape function deals with the flux of energy, information, and materials within and among the component ecosystems (elements) forming the landscape. Nowhere in landscape ecology is the interdisciplinary nature of the enterprise more evident than in the study of landscape function. Various kinds of environmental scientists (hydrologists, geomorphologists, meteorologists, pedologists, statisticians, ecologists, etc.) and domain specialists (foresters, entomologists, agronomists, landscape architects, etc.) typically are needed to decipher the mechanisms and consequences of landscape function.

In Chapter 1 we introduced the pattern/process paradigm, the essence of which is that characteristic patterns in landscapes result from the operation of ecological processes and that by studying these patterns we can make useful inferences about the underlying processes[1]. The reverse of this statement is true as well: by studying ecological processes we can make useful inferences about observed patterns in landscapes. To a considerable extent, the ecological processes are actually movements or flows of energy, information, and abiotic materials into, within, and out of the landscape. In the introduction to their text, Reiners and Driese (2004) posited the fundamental question that underpins the subject of

---
[1] Herein, a *process* is a defined simply to be a change that leads towards a predictable result (Kelmelis 1998).

landscape function: "How do the events or conditions in one area of an environmental domain lead to the transmission of effects across that domain to produce an ecological consequence somewhere else?" In this chapter we have organized our discussion of landscape function in the context of this question. The goal of this chapter is to examine landscape function from a mechanistic perspective. The discussion is divided into five sections. In the first section, we present and review general tenets of a model of landscape function. The next three sections deal with the basic elements of the model. In section two, we examine the causal agents that initiate the propagation of ecological effects. In section three, we consider the nature of the entities that are propagated across landscape space, i.e., energetic entities, information entities, and material entities. In section four, we consider how the ecological entities are propagated, i.e., the means of transport of the entities. In section five, we examine the consequences of propagation of ecological effects (Figure 8.1).

GENERAL MODEL OF LANDSCAPE FUNCTION

The transport model, illustrated in Figure 8.2, was developed from an environmental science perspective. With minor constraint, the model is directly applicable to our examination of landscape function. In this section we consider two issues associated with the use of the model in the context of landscape ecology. First, we provide a general overview of the

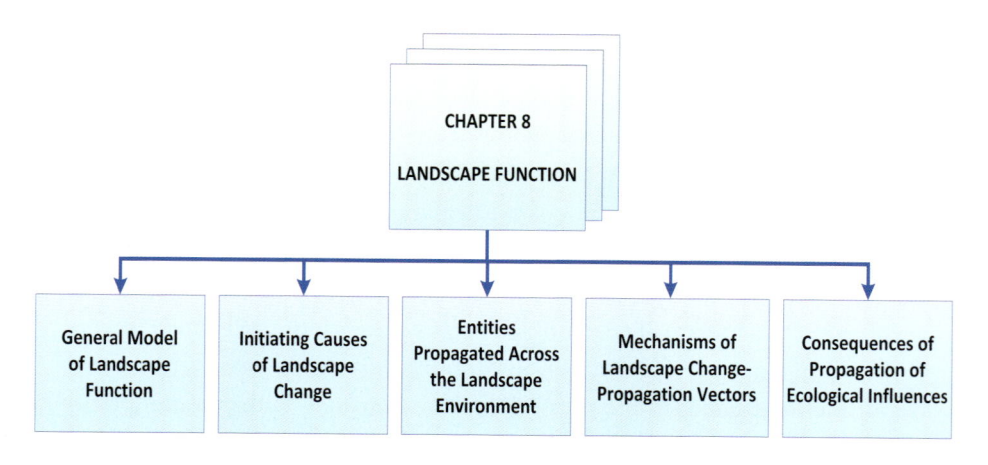

Figure 8.1 — Summary of topics and organization of Chapter 8.

model and introduce the basic vocabulary needed to understand how it operates. Second, we consider the application of model within the spatial and temporal boundaries of landscapes.

Model Overview

In their text, *Transport Processes in Nature,* Reiners and Driese (2004) provided a novel perspective and general model that addressed the "propagation of ecological influences through environmental space[2]." Figure 8.2 is a tailored rendition of this model. With some concession to simplification, we have adapted this model to frame the following

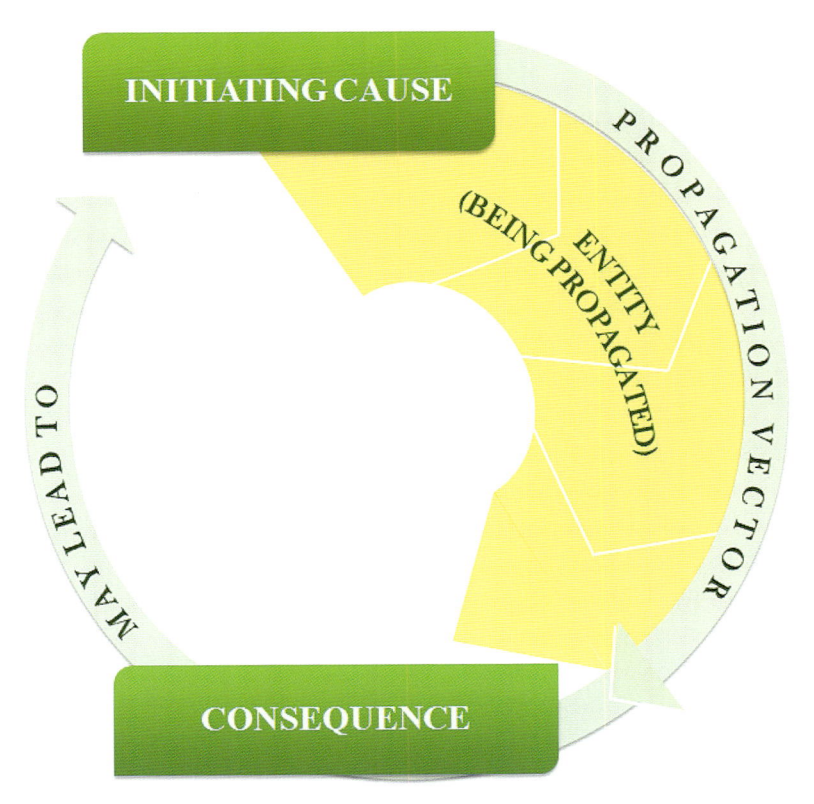

Figure 8.2 —Diagram representing of the general concept of propagation of ecological effects across the landscape environment. There are four basic components to this model of transport: an initiating cause, an entity, a propagation vector, and a consequence (modified from Reiners and Driese 2004).

---

[2]What does this statement mean to you? The authors carefully selected these words. Their intended meaning will become evident as you progress through this chapter.

Landscape Function | 161

discussion of landscape function. The model is a metaphor (heuristic theory) of landscape function. There are four basic components: an initiating cause, an entity, a propagation vector, and a consequence (Table 8.1). The *initiating cause* is defined as the factor that brings about an effect or a result. Examples of initiating causes include atmospheric disturbances (thunderstorms, hurricanes, tornados), fire, herbivory, pathogen infection, etc. A cause can be characterized by a number of different descriptors: e.g., it can be biotic or abiotic in origin; it can be distributed, targeted, diffuse, or patchy in space; it can be frequent, rare, or periodic in occurrence; etc. The initiating cause leads to the movement of some entity (dust, sound, gases, spores, etc.) from its origin to a point of deposition in another place in the landscape. The *entities* (i.e., what is transported) can be classed generally as energy, information, and matter. Specific entities occur in a variety of forms, e.g., photochemical oxidants, pollen grains, pathogen spores, insects, lightning, etc. Typically, entities are bundled, i.e., a pollen grain consists of information, energy, and matter. The *propagation vector* is the means for transport or conveyance of the entity. Synonyms are *carrier* or *conveyor*. There are several types of propagation vectors, and they can be classified in a variety of ways (Table 8.1), e.g., fluvial (water),

Table 8.1 — Basic vocabulary for describing landscape function.

1.	*Initiating cause*	A factor that brings about an effect or a result, e.g., a fire, hurricane, herbivory, introduction of a pathogen, etc.
2.	*Entity*	Something that has separate and distinct existence and objective or conceptual reality. Entities represent what is being propagated or transported within the landscape. Entities can be classed as matter, information, and energy. The entities include many specific forms, e.g., dust particle, pollen grain, water, herbivores, etc.
3.	*Propagation vector*	The means for transport or conveyance of the entity, e.g., wind, water, gravity, animal locomotion, etc. The propagation vectors can be classified in different ways, and in the text we examine a range of possibilities.
4.	*Consequence*	The effect(s) resulting from the propagation of an entity from one place to another in a landscape. Examples of consequences are soil erosion, disease epidemics, and pollination of flowering plants.

aeolian (wind), mass movement, animal movement, etc. The effect of vector-mediated movement of an entity from one place to another within the landscape is referred to as a *consequence* (the right arrow in Figure 8.2). Furthermore, the consequence may become another initiating cause that leads to secondary or tertiary effects (the left arrow in Figure 8.2). In some instances the consequence of transporting an entity can have a significant impact on landscape structure. Therefore, we consider landscape change as a separate topic in Chapter 9. Table 8.2 provides several examples that illustrate the relation of initiating phenomena, entity transmitted, vector transmitting the entity, and consequences of transmission in landscapes. The steps in the transport model are straightforward: an initiating cause, the vector-propagation of a specific entity, and a resulting consequence. However, the secondary and perhaps tertiary effects may not be obvious. This subject is discussed in more detail later in the chapter.

Table 8.2 — Examples of ecological influences transmitted across landscape space defined by initiating causes, entities, vectors, and consequences (modified from Reiners and Driese 2004).

Initiating Phenomenon on Part of a Landscape	Entity Transmitted Across the Landscape	Vector Transmitting the Entity	Consequence of Transmission to Another Part of the Landscape
Volcano eruption	Ash	Wind	Fish mortality in streams and lakes
Malaria outbreak	Disease microorganism (*Plasmodium*)	Mosquito	Mortality to humans
Rainstorm	Water	Fluvial transport driven by gravity	Erosion, nutrient transport, etc.
Thunderstorm	Lightning	Electromagnetic radiation	Combustion, herbivory
Flowering of plants	Pollen	Wind, animal locomotion	Plant reproduction
Bird "song"	Sound	Sound waves	Mate attraction

Application of the Model to Landscape Ecology

As the transport model was developed from the perspective of environmental science, its application is not constrained by spatial and temporal scales. Terms such as *environmental domain* and *environmental space*, which were casually introduced in the preceding description, accommodate local to global space and time scales. However, throughout this text we have been vigilant in our use of the term *landscape*: a spatially explicit geographic area, i.e., an area defined by coordinates, consisting of recognizable and characteristic component entities (ecosystems). Furthermore, we have emphasized that the extent (size) of the landscape is a unit with flexible spatial dimensions, which can be specified for practical purposes. So in applying the transport model, in the context of *landscape function*, the major intra- or inter-landscape variable is the spatial (and temporal) scale of the propagation vectors. For example, when Mount St. Helens erupted (Figure 5.9) the event occurred within a discrete landscape. One effect of the eruption was production of immense quantities of volcanic ash (Figures 8.3a and b). This entity was transported from the local landscape (where the event initiated) by wind currents (the propagation vector) to surrounding landscapes and ecoregions. Indeed, within 15 days the ash was distributed globally. The consequences of transporting this entity are manifold, but one direct effect was on native freshwater and anadromous fish species (Bisson et al. 2005). In addition to direct mortality to the fish resulting from debris avalanches and mudflows; elevated concentrations of volcanic sediment, physical destruction of breeding habitat, and change in chemical properties of streams greatly affected re-establishment and reproduction of the anadromous fish species in particular. Stream ecosystems in landscapes far removed from the site of the eruption event were affected, as a consequence of wind transport of the ash. In this example, the initiating cause and the consequence are site specific, i.e., they occur in discrete ecosystems of affected landscapes. However, the spatial scale (extent) over which the propagation vector operated was extensive. Propagation vectors certainly can operate locally within the ecosystems of a specified landscape, e.g., a squirrel (*Sciurus* spp.) collecting, moving, and cashing nuts in an oak/hickory forest. The initiating cause of a change process and the resulting direct consequence are site-specific, but the spatial extent of the transport vector is a variable.

Figure 8.3a — The eruption of Mount St. Helens produced immense quantities of volcanic ash. This entity was vectored by wind and impacted native and anadromous fish populations in stream and lake ecosystems (USGS photograph).

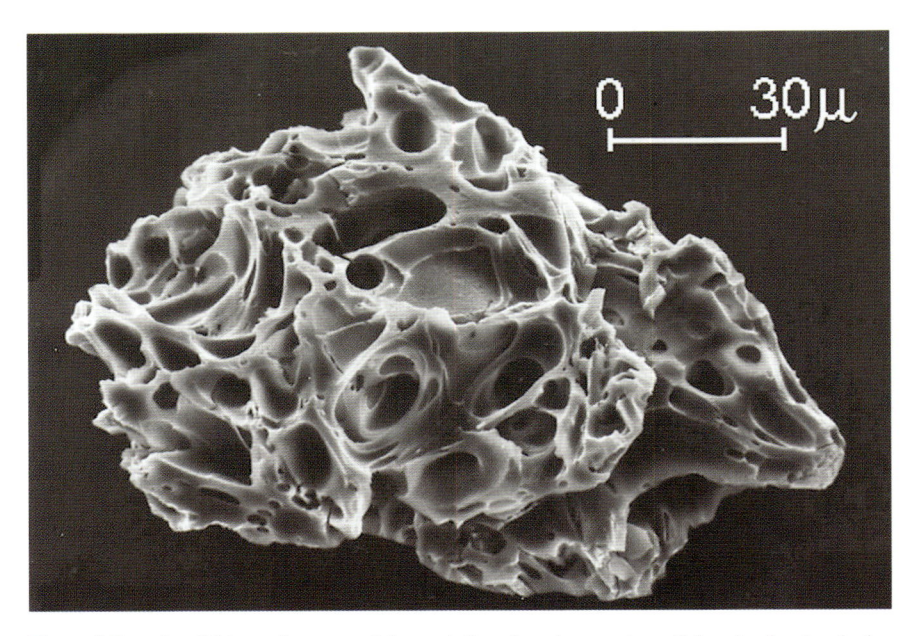

Figure 8.3b — In addition to impact on fish population, the microscopic particles of volcanic ash also affected human respiratory health (USGS photograph).

Landscape Function | 165

ATTRIBUTES OF CAUSAL AGENTS THAT INITIATE THE PROPAGATION OF ECOLOGICAL EFFECTS

Recall that an *initiating cause* is defined to be a factor that brings about an effect or a result. Landscapes are replete with various kinds of causal agents, e.g., a rain storm, a volcano eruption, an insect outbreak, a forest fire, etc. Although the possibilities for different kinds of initiating causes are immense, there are several general descriptors that can be applied to catalog them. The descriptors are useful for organization and characterization purposes. The basic descriptors for initiating causes are presented below as questions. A landscape ecologist could provide answers to the questions. The answers facilitate comparisons of different initiating agents. The questions are:

- Is the initiating cause biotic or abiotic in origin?
- Does the initiating cause originate within or outside of the landscape?
- Is the initiating cause natural (derived from nature) or an anthropogenic (derived from humans)?
- Is the initiating cause a discrete event or a chronic condition?
- What is the timing of the initiating cause?
- What is the scale (spatial extent) over which the initiating cause operates?

Following is a brief explanation of each of the descriptors. The initiating causes can and do operate over very large spatial and temporal scales. Our interest here is focused on initiating causes associated with the ecosystem components of landscapes.

Biotic or abiotic: The initiating cause is a living organism (e.g., a plant pathogen) or a non-living physical force (e.g., a windstorm).

Origin within or from outside the landscape: An initiating cause that comes from within the landscape is referred to endogenous (e.g., an insect outbreak in a forest landscape). An initiating cause that comes from outside the landscape is referred to as exogenous (e.g., a lightning strike).

Natural vs. anthropogenic: A natural initiating cause derives from nature. It can be biotic (living) or abiotic (physical). A bird "song" is an example of a natural initiating cause. An anthropogenic initiating cause derives from human activity (e.g., the sounds created by motor vehicles on a road corridor).

Discrete or chronic event: A discrete initiating cause is a distinct incident (e.g., an insecticide application to a cotton field to suppress pest insects). A chronic initiating cause is characterized by long duration or frequent recurrence. An example is sulfur dioxide smoke-stack emission from a coal-fueled electrical power generation plant.

Timing of the initiating cause: Timing deals with frequency of occurrence of an initiating cause. The relevance of timing is usually judged in reference to the effect on specific organisms and is scaled in terms of daily, monthly, seasonal, annual, etc. cycles. Timing of initiating causes also has a spatial component associated with it[3], e.g., how frequently does an initiating cause occur at a specific location.

Scale of the initiating cause: Scale refers to the spatial extent of the initiating cause. Initiating causes can be targeted to discrete ecosystems within the landscape (e.g., a forest fire) or encompass multiple ecosystems (e.g., a tornado). An initiating event such as an earthquake can affect multiple landscapes.

ENTITIES PROPAGATED ACROSS THE LANDSCAPE ENVIRONMENT

In order for an initiating cause originating at one place in a landscape to have an effect at another place, something discrete (an entity) must be transferred between the two locations. As with initiating causes, entities exist in many specific forms, but all can be conveniently classed as energy, information, or materials. In this section we examine each type of entity separately. As entities are typically combinations of energy, information, and materials; we conclude with a discussion of this interrelatedness.

---

[3] How does "Hopkins' Bioclimatic Law" relate spatial and temporal components of periodic biological phenomena?

The Entities

Although the terms *energy*, *information*, and *materials* are used in common parlance, their meaning, in the context of landscape function, requires elaboration. Our task here is to clarify and specify what these terms mean in the context of landscape function. Later in this chapter, we will consider how the entities move. In Chapter 7 we considered energy and materials in the context of resources needed by living organisms.

Energy — Energy is a fundamental property of material and living systems. It is defined simply as the capacity to do work or to produce change. As an entity, energy is a *scalar* in mathematical terms, i.e., a quantity that can be fully described by a magnitude or numerical value alone. Earlier, in this chapter we defined a process to be a change that leads to a predictable result. By this definition, energy exchanges or transfers are associated with all processes, regardless of whether they occur in physical or living systems. Energy transfer in a system is manifested as either work or heat. Energy exists in a variety of forms and can be converted between each type. Table 8.3 provides a systematic classification based on potential energy (energy of position) and kinetic energy (energy of movement).

In landscape ecology our interest in energy is generally directed to issues associated with processes of geomorphology and ecological energetics. In particular, geomorphology is concerned with energy associated with the basic driving forces of climate (e.g., wind and ocean currents), gravity-driven processes (e.g., the basic geomorphological processes – alluvial, fluvial, mass movement, coastal, etc.), and nuclear reactions (e.g., radioactive decay, friction by rock deformation) (Ritter et al. 2002). Ecological energetics traditionally has dealt with the flow of energy through the medium of living organisms, i.e., food chains and food webs (refer to Chapter 7). We are also concerned with energy associated with movement of organisms. In both the physical and biological systems, the flow of energy is governed by the first and second laws of thermodynamics.

Information — The landscape ecological perspective of information as a transport entity requires that we consider the meaning of this word in the context of a three-level hierarchy (Figure 2.7) that includes data, information, and knowledge. In this hierarchy *data* (the bottom level)

Table 8.3 — Classification of the basic forms of energy.

Kinetic Energy -- Motion	Potential Energy -- Position

• Mechanical energy:
    ▪ Machines
    ▪ Wind energy
    ▪ Wave energy
    ▪ Sound (sonic, acoustic) energy
• Thermal energy:
    ▪ Geothermal energy
• Electrical energy:
    ▪ Lightning
• Electromagnetic radiation:
    ▪ Radio, microwave, infrared, light, ultraviolet, x-rays, gamma rays
    ▪ Solar energy
	• Gravitational potential energy
	• Electromagnetic potential energy
	    ▪ Electric potential energy
	    ▪ Magnetic potential energy
	    ▪ Chemical potential energy
	    ▪ Elastic potential energy
	• Nuclear potential energy
	    ▪ Nuclear power
	    ▪ Radioactive decay

are the measurements that define an ecological phenomenon, process, or relationship of interest. For example, when an ornithologist records a bird "song" in the forest, the data captured in this activity is sound, represented as a series of compression waves. The actual data are properties of the waves: frequency, wavelength, period, amplitude, intensity, speed, and direction. The second level in the hierarchy is *information* (the center level of focus or interest), which is defined to be data that have been given meaning by way of relational connection. When the ornithologist can attribute a "song" to a particular bird species, the sonogram recording becomes information. The third level in the hierarchy is *knowledge* (the top level), which is defined to be contextually integrated information, i.e., knowledge consists of an organized body of information. For example, upon further study of sonogram recordings, coupled with observations in the forest landscape, the ornithologist recognizes that the bird species she has been studying has several different "songs" and that each has a unique function, e.g., one song is used to identify a territorial boundary, another to warn of predators, and yet another to attract a mate. The contextual integration of this information provides the ornithologist with insight into the behavior of the bird. The integrated information represents knowledge. The data/information/knowledge hierarchy is not simply a semantic construct. We will use this hierarchy again in Chapter 10 in the discussion of methodologies for analysis, synthesis, and application of spatial data, information, and knowledge.

The landscape ecological significance of data, information, and knowledge involves consideration of these entities from two perspectives. The first deals with behavioral interaction of organisms in the landscapes. The details of communication among many organisms are guided by perception, integration, and response to data, information, and knowledge of the landscape environment in which they live. The second perspective deals with genetically encoded "information" represented as DNA. Spores, pollen, seeds are examples of entities where "information" has been encapsulated in a form that can be vectored throughout landscapes and beyond. The transport mechanisms and the consequences of movement of "information" packaged in this form are different from the behavioral interactions associated with communication among organisms.

In the description of DNA encryption (associated with spores, pollen, seeds, etc.), we bundled data, information, and knowledge under the general heading of "information," hence our apologetic use of quotations. However, these packages represent the evolutionary history of the organism. What do you think: Do the propagules represent data, information, and knowledge? Recall that information is the central element or focal point of our hierarchy and that data (the lower level) is used for explanation and knowledge (the upper level) for interpretation (see Chapter 2). A unique feature of the "information" entity is that it must be encoded in either an energetic or material context, regardless of whether represented as behavioral signals or genetic code, i.e., unlike energy and matter, information does not have a pure form.

Materials — *Materials* are defined broadly to be the elements, constituents, or substances of which something is composed or can be made. The global perspective of materials in the environment is the subject domain of physical geography and includes consideration of substances such as the minerals associated with the lithosphere, hydrosphere, and atmosphere.

Abundant minerals found in lithosphere include the following: O, Si, Al, Fe, Ca, Mg, Na, K, Ti, H, Mn, and P.

The landscape ecological perspective on materials is addressed in concepts associated with *nutrient cycling*, which is defined as the transformation of chemical elements from inorganic form in the environment to organic form in organisms, and via decomposition back into inorganic form. The general process is illustrated in Figure 4.5. Nutrient cycles involve two general phases: the *organismal phase*, in which the nutrients exist in organic form as the living tissues of organisms; and an *environmental phase*, in which the chemical nutrients are in inorganic form and occur in the soil, water, or air. The materials of the organic phase are principally water (H and O) and C compounds. The materials of the environmental phase include N, P, K, S, Ca, Mg, Fe, and Na. Energy flow and nutrient cycling in landscapes are inseparable.

The Interrelatedness of the Transport Entities

In the previous section we examined energy, information, and materials as independent entities. The domains of study for energy and materials are the basic sciences of mathematics, physics, and chemistry. Information is not as cleanly compartmentalized and includes topics such as epistemology, information theory, informatics, and knowledge engineering. These last subjects are addressed by a variety of disciplines ranging through philosophy, psychology, computer science, and engineering. However, as the agents that actually impact on the landscape environment, the entities are frequently integrated combinations of energy, information, and matter. An understanding of the basic nature of each type of entity is a necessary prelude to the examination of the interactions. Energy and matter have pure forms, e.g., abiotically generated radiation and sound, and non-reactive inorganic materials. As noted above, information by definition is contextual and therefore does not exist independently. However, the assignment of an entity to a particular compartment requires consideration of its effect (consequence) in the landscape (Reiners and Driese 2004).

VECTORS RESPONSIBLE FOR TRANSMISSION OF ENTITIES IN THE LANDSCAPE ENVIRONMENT

In contrast to scalars, *vectors* (in mathematical terms) are quantities which are described by both magnitude and direction. Vector quantities include displacement, velocity, direction, force, momentum, weight, drag, thrust, acceleration, and lift. In keeping with our emphasis on the mechanism of landscape function, the intent in this section is to systemize and define the various types of vectors. The operation of vectors in transporting entities results in consequences, often manifested as landscape change, and this topic is investigated in Chapter 9.

For discussion purposes we can organize vectors that transport ecological entities into the following three compartments: processes of geomorphology, movement of humans and their use of transport technologies, and movement of living organisms (other than humans). However, this organization scheme is for discussion purposes, as the transport vectors seldom operate in isolation.

Processes of Geomorphology

Geomorphology is the science of landforms, including their history and origin. In physical geography, *landforms* are configurations of the land surface having distinctive character and produced by natural processes. A landform includes surface geometry and underlying geologic material. The terms *physiography* and landform are synonyms (Bailey 2002). Often, the landform is defined by the process responsible for its formation, e.g., depositional, erosional, tectonic, etc. The basic geomorphology processes that serve as vectors of energy, information, and materials include the following: aeolian, fluvial, mass movement, and coastal. Weathering and glacial processes are important in landform formation. However, as weathering occurs *in situ*, i.e., without movement, the process is not directly involved in vector transport of entities within landscapes. Glaciers certainly act as vectors for movement of entities, but the temporal scale (rate of movement) is generally not relevant to landscape ecological investigations (Figure 8.4). Here, we simply define the geomorphology processes in the context of their role as vectors that move entities consequential to landscapes. Summerfield (1991) provides a thorough and

Figure 8.4 — Glaciers act as vectors for movement of entities, but the temporal scale (rate of movement) is generally not relevant to landscape ecological investigations. However, the landforms created by past glacial events are certainly an important component of landscape structure (British Columbia, Canada) (KEL image).

scholarly examination of the processes. Later in the chapter we examine the relation of landform and landscape structure.

Aeolian Processes — In a landscape ecology context, *aeolian processes* deal with wind transport of entities (Figure 7.16). The geomorphology focus centers on sediment transport. There are three basic components to movement of sediments: entrainment of entities, vertical and horizontal movement by wind currents, and deposition and erosion. Although details of the mechanisms are different, aeolian-transported sediment moves in three basic ways: in suspension, by saltation (bouncing), or by surface creep (Figure 8.5).

The other vector function provided by wind is transport of living organisms in the atmosphere. The flow of biota in the atmosphere is a function

Landscape Function | 173

of biological attributes of living organisms as well as meteorological conditions. Together this interaction is the substance of the study of aerobiology (Isard and Gage 2001). Many organisms move within and among habitats in terrestrial landscapes by floating, soaring, and flying. These modes of movement require much less energy per unit body mass than walking. Isard and Gage (2001) emphasize that there is more movement among terrestrial habitats by organisms that use the air than by those that move over land surface or through water. Later in this chapter we examine the role that living organisms play as vectors for movement.

Fluvial Processes — In a landscape ecology context, *fluvial processes*, deal with water transport of entities (Poole 2002). In terrestrial landscapes, the processes take place at the scale (spatial extent) of entire drainage basins. This scale is a variable and can range from sites convenient to landscape ecological investigations (e.g., the Coweeta Basin illustrated in Figures 2.18, 2.19, and 2.20) to areas that encompass drainages of major river systems (e.g., the Mississippi River basin). The vector function of water is tied to the basic hydrologic process, which is summarized in Figure 8.6. The entities associated with water movement include materials associated with the drainage basins, dissolved and undissolved chemical compounds, and living organisms. Movement of entities takes place by rolling, sliding, saltation, and suspension.

Mass Movement — *Mass movement* is defined as gravity-driven downward movement of slope material without the assistance of moving water, ice, or air (Summerfield 1991). *Mass wasting* and *colluvial transport* are synonyms (Figure 8.7). Slopes are a basic element of landscape topography, and therefore, have been a central focus for landform studies. Downslope movement of materials under gravitational force is influenced by a number of variables including constituency (rocks, soil), frictional resistance between component particles of the material, slope steepness, water content, etc.

Coastal Processes — *Coastal processes* include the effects of waves, tides, and currents occurring at the interface between the terrestrial and marine (or other large water body) environments. This interface, *the littoral zone*, has two components: a shallow water zone and a terrestrial zone. Waves, tides, and currents act as transport processes and serve to move entities

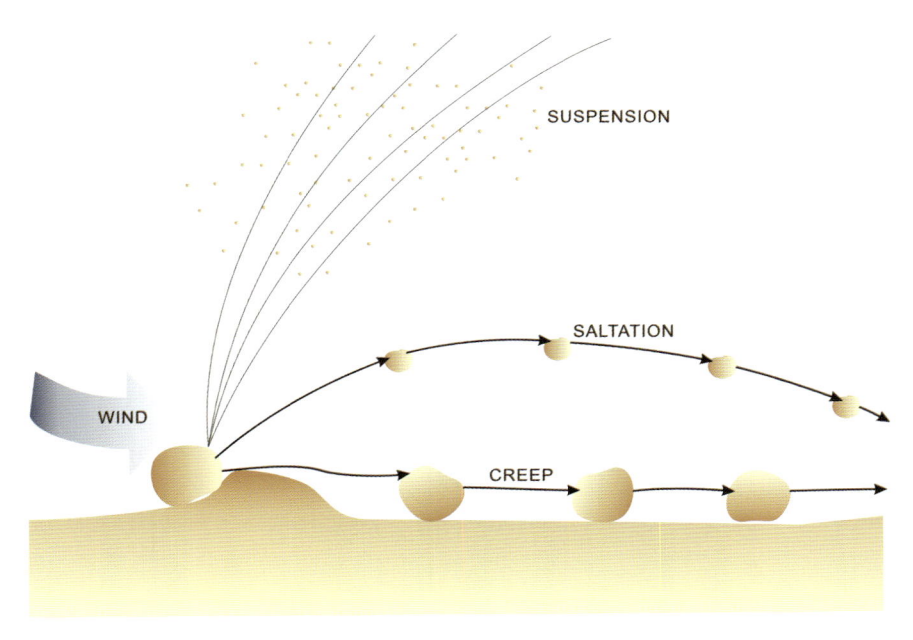

Figure 8.5— Aeolian entity movement takes place in suspension and by saltation and creep (modified from Summerfield 1991).

within the shallow water zone. The terrestrial zone includes beaches, cliffs, coastal dunes, and lagoons. These landforms are also affected to some degree by the transport processes as well. Coastal processes affect the littoral zone in two ways: by destruction and construction. *Destructional processes* involve shoreline weathering and coastal erosion (Figure 8.8). *Constructional processes* involve material movement and deposition. Noteworthy examples of constructional processes include the vectored movement of seeds that lead to the establishment of mangrove forests (Figure 8.9 a and b) and nutrients and organisms involved in coral reef development.

Humans as Transport Vectors in Landscapes

Humans serve as transport vectors for energy, materials, and information in the landscape environment. The vector function is accomplished in three principal ways: through locomotion (movement from place to place under human power), as carriers of other organisms, and through the use of transport technologies. Mechanized landscape domestication practices

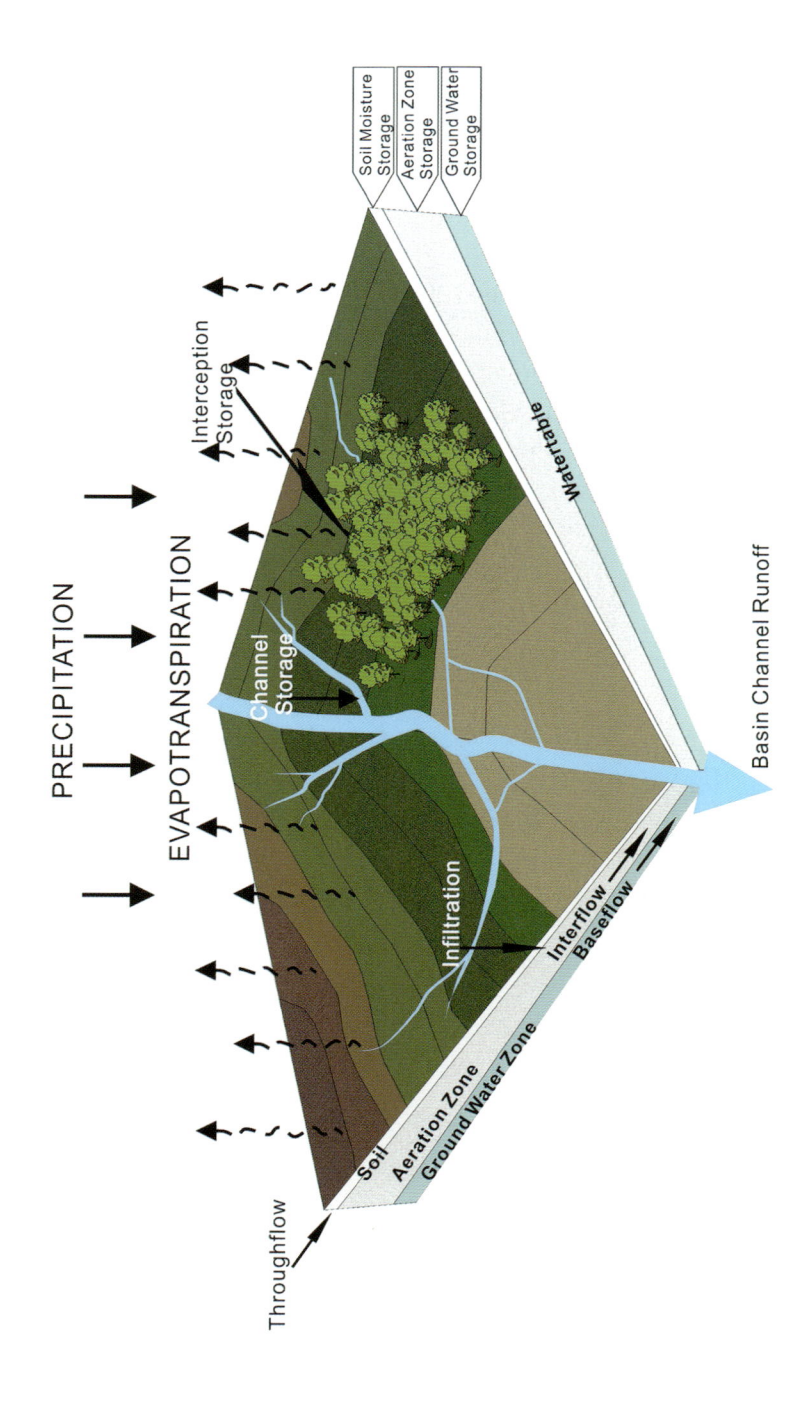

Figure 8.6— The hydrologic cycle summarizes water movement in landscapes. The general scope of hydrology is concerned with water input, output, and storage within the basin. All water enters the drainage as precipitation. It can be stored in the soil, aeration zone, or ground water zone. Driven by gravity, water moves out of the basin and is eventually returned to the atmosphere by transpiration (KEL drawing).

176 | *Landscape Function*

are intended, in part, to reduce and optimize actual human participation in vector transport. Movement of the entities is directly associated with activities involved in or a consequence of landscape domestication (Kareiva et al. 2007).

Living Organisms (other than humans) as Transport Vectors in Landscapes

Living organisms, in addition to humans, also act as transport vectors of energy, information, and materials in the landscape environment. In particular, many animals move about in the landscape environment by various means of locomotion. Movement in space takes place across terrestrial landscapes, within the air, and through water. Adaptation in morphology, physiology, and behavior establish the specific mode of movement utilized by different species, e.g., walking, running, jumping, slithering, burrowing, flying, swimming, etc. (Reiners and Driese 2004). Details of these adaptations will also establish capacity for movement, i.e., how far and how fast an animal can move.

Animals, as well as plants, also move as passengers on and in other animals. Various types of symbiotic relations result in movement of associated organisms in the landscape environment. The associations, which are the substance of community ecology (see Figure 2.5), can range from simple phoresy (hitching a ride) to obligatory parasitic relationships that result in movement and transmission of disease organisms. Indeed, disease epidemiology integrates knowledge of the interaction of the animal vector, disease, and host organism in the context of the landscape environment.

CONSEQUENCES OF PROPAGATION OF ECOLOGICAL EFFECTS IN THE LANDSCAPE ENVIRONMENT

In working through the components of the transport model (Figure 8.2), we have examined basic issues associated with initiating causes, entities, and movement vectors. In this section we consider the remaining element of the model: the consequences of moving entities from one place to another in the landscape environment. Two issues are of particular importance: interpretation of cause and effect scenarios and ecological effects as episodic events (Figure 8.10).

Figure 8.7 — Landslides are a kind of mass movement. This image illustrates a landslide and earth flow that occurred in the spring of 1995 in La Conchita, California. People were evacuated, and the houses nearest the slide were completely destroyed (photograph by R.L. Schuster, USGS).

Figure 8.8 — Coastal processes include the effects of waves, tides, and currents occurring at the interface between the terrestrial and marine (or other large water body) environments. The Punakaiki (Pancake Rocks Blowholes) in New Zealand illustrate destructional coastal weathering (KEL image).

Figure 8.9— Coastal process can also be constructional. The mangrove forest (a) resulted from the tidal transport and subsequent establishment of seeds (b) (Galapagos Islands, Ecuador) (a, KEL image, b, NPS photograph).

Interpretation of Cause and Effect Scenarios

The transport model (Figure 8.2) is a mechanistic construct that sequences cause and effect, i.e., it addresses *how* an initiating cause, an entity, a propagation vector, and an effect (consequence) are related dynamically. Investigations centered on how ecological systems function (or operate) are commonplace in ecology. "How" questions fuel the research agenda for many landscape ecologists. However, landscape ecologists and land-use managers are also interested in questions that center on *why* ecological systems operate as they do. "Why" questions are not dynamic. Answers are typically based on integration of multiple factors, i.e., they involve the examination and interpretation of an ensemble of events.

Our focus centers on how the transport model deals with cause and effect. In some instances the result of a specific cause and effect scenario differs as a function of interpretation by an observer. In other cases, the same cause and effect scenario leads to a fundamentally different result. The following examples illustrate these two situations.

First, in the example presented above involving the transport of volcanic ash from the Mount St. Helens volcano, the effects (consequences) were interpreted from a fisheries biologist perspective. A physician in an urban landscape would likely be more concerned with the effects of volcanic ash on respiratory illness of human inhabitants. Specific interpretations of cause and effect scenarios are often relative to the domain interest of human observers (Figure 8.10).

Second, a cause and effect scenario can produce fundamentally different results. For example, consider the circumstance where an atmospheric disturbance produces lightning. Multiple consequences can follow. If the lightning strikes a tree in a pine forest landscape in the southern United States, ignition can follow and result in a forest fire. However, in some instances the lightning striking a pine tree only produces a wound face along the surface of the bole. A tree in this condition is often colonized by phloem boring bark beetles (Coleoptera: Curculionidae), and an infestation consisting of multiple trees subsequently develops (Lovelady et al. 1991). In both cases the age structure and species composition of the forest landscape are changed. However, in the case of the fire event, nutrients

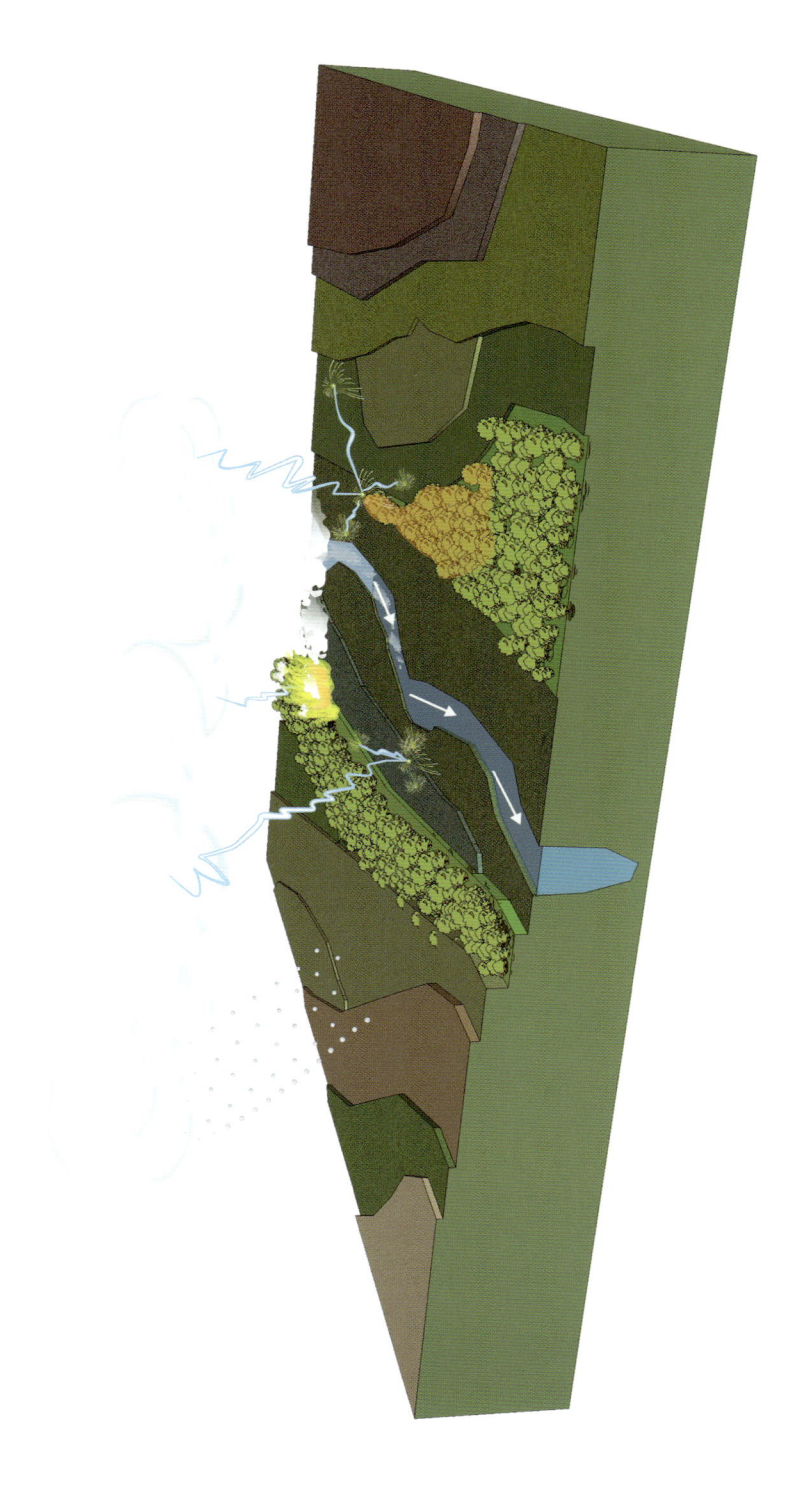

Figure 8.10 — The steps in the transport model are straightforward: an initiating cause, the vector-propagation of a specific entity, and a resulting consequence. Focus can be, and typically is, directed to a specific cause/effect relation. However, a specific cause and effect episode is likely to be a member of a chain of sequenced or related events. In this figure lightning striking a tree can ignite a forest fire or in some instances trigger an infestation of phloem-boring insects. The forest fire produces gas and debris (a secondary episode), which is transported to ecosystems in adjacent landscapes (producing a tertiary episode) (KEL drawing).

182 | *Landscape Function*

are exported by wind currents from the landscape as smoke and debris. In the case of the insect outbreaks, nutrients are retained within the landscape and are recycled through decomposition processes. Two different effects result from the same initiating cause (Figure 8.10).

Consequences as Episodes

The transport model accommodates cause and effect relations as a feedback process. This relation is indicated by the left side arrow in Figure 8.2. Causes lead to effects, which trigger new causes, which lead to new effects. For example, lightning (a cause) strikes a tree in the forest and ignites it (an effect), the ignited tree (a cause) leads to a forest fire (an effect), the forest fire (a cause) produces gases and debris which pollute adjacent landscapes (an effect), etc. Focus can be, and typically is, directed to a specific cause/effect relation. However, a specific cause and effect episode is likely to be a member of a chain of sequenced or related events. An *episode* is an event that is distinctive and separate although a part of a larger series, i.e., a situation that is integral to but conceptually separable from a continuous process or stream of events (Rykiel et al. 1988).

EPILOGUE TO LANDSCAPE FUNCTION

So what conclusions can be made about landscape function?

1. Landscape function deals with the flux of energy, information, and materials within and among the component ecosystems forming the landscape. The discussion of landscape function was organized to address the following fundamental question: "How do the events or conditions in one area of an environmental domain lead to the transmission of effects across that domain to produce an ecological consequence somewhere else" (Reiners and Driese 2004).

2. A general transport model, consisting of four components, was used to examine landscape function from a mechanistic perspective. The dynamically related components include an *initiating cause* (the factor that brings about an effect), an *entity* (what is transported), a *propagation vector* (the means of transport), and a *consequence* (the effect of the vector-mediated movement of the entity on the landscape

environment). The transport model accommodates cause and effect relations as a feedback process, i.e., causes lead to effects, which trigger new causes, which lead to new effects.

3. The initiating cause of a change process and the resulting direct consequence are site-specific within a landscape, but the spatial extent over which the transport vectors operate is a variable and can include long distances.

4. Each element of the transport model was examined systematically. Initiating causes were characterized by general descriptors that are useful for organization and characterization purposes. The entities transported included energy, information, and materials. The entities were described separately. However, emphasis was placed on the fact that the agents that actually impact the landscape environment are frequently integrated combinations of energy, information, and matter. The transport vectors of ecological entities included the basic geomorphology processes (aeolian, fluvial, mass movement, and coastal), human-mediated activities associated with landscape domestication, and activities of organisms (other than humans). Discussion of the consequences of moving entities from one place to another in the landscape environment included an examination of the nature of cause and effect relationships, interpretation of cause and effect scenarios, and ecological effects as episodic events.

9

Landscape Change

OVERVIEW

Landscape change deals with the alteration of structure and function of the landscape environment over time. In Chapter 7, we examined landscape structure, i.e., components of the landscape and their linkages and configuration, from three perspectives: environment, geometry, and perception. Major emphasis was placed on geometry, as the specific ensemble of components that make up the landscape mosaic represent the stage upon which the drama of ecology is played out. In Chapter 8 we examined landscape function, i.e., the flux of energy, materials, and information within the landscape environment, from a mechanistic perspective. The transport model (Figure 8.2) was used to illustrate how an initiating cause, an entity, a propagation vector, and an effect (consequence) are related dynamically. In general, the effect or consequence of vector transport of an entity results in a change in the landscape. Although the vectors can operate over large areas (spatial extent), the effects that are consequential to our study of landscape ecology are site-specific. The concepts of landscape structure and landscape function are inextricably tied to one another. By analogy – pattern : process :: structure : function.

In this chapter our goal is to consider the causes and consequences of landscape change. We have four objectives. The first objective deals with landscape-cover change. In particular, we examine the concept of landform and biogeomorphology, the roles that living organisms (other than humans) play in changing landscapes, and the concept of ecological disturbance. The second objective deals with landscape-use change. In particular we examine landscape-use change in the context of landscape domestication. We consider how human needs lead to landscape-use change and introduce the concept of drivers of change. The third objective deals with how landscape change affects the ecology of living organisms. We examine landscape change in the context of habitat modification. This discussion centers on how landscape change affects the persistence, distribution, and abundance of living organisms. The fourth objective deals with the development of pattern in mosaic landscapes. We examine how the integration of processes of landscape-cover and landscape-use change results in observed patterns of landscape mosaics (Figure 9.1). The transport model (Figure 8.2) serves as the preface to this discussion of change. The various mechanisms of change, discussed in the following sections, are simply types of initiating causes, and the consequences of their operation result in alteration of landscape structure and function.

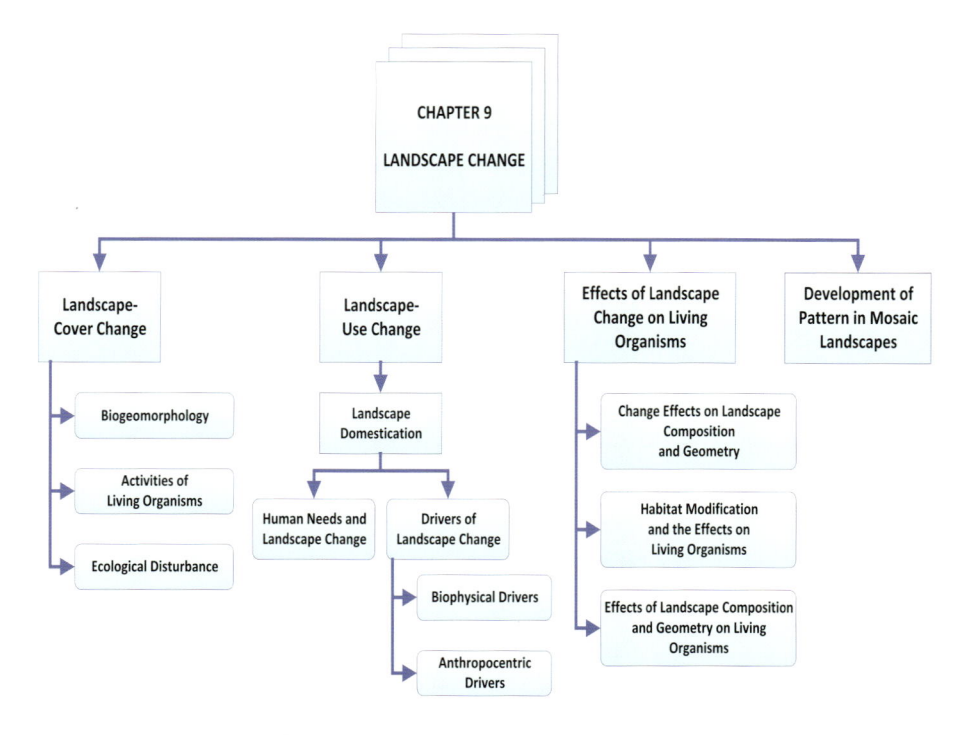

Figure 9.1 — Summary of topics and organization of Chapter 9.

The contemporary literature in environmental science and management is laden with reference to land-use/land-cover (LULC) change, often in the context of global climate change, land degradation, deforestation, desertification, loss of biodiversity, habitat destruction, etc. We have tailored these terms for our discussion of landscape change: *landscape-use* and *landscape-cover*. The rational for this constraint is to retain a focus on landscape as place-based concept, i.e., a spatially explicit geographical area consisting of characteristic component entities (ecosystems). This point of emphasis underpins the following examination of landscape change.

The subjects of landscape-cover change and landscape-use change are highly interrelated. At the onset of this text we acknowledged that humans and their actions were, *de facto,* an element of landscape ecology. A great deal of discussion in landscape ecology centers on the responses of the community of life to physical and human-induced changes in the landscape environment.

LANDSCAPE-COVER CHANGE

Landscape-cover change deals specifically with the alteration of biophysical attributes of the landscape environment (Lambin et al. 2001). In the following sections, three aspects of this subject are examined: the concept of landform and biogeomorphology; the roles that living organisms (other than humans) play as ecosystem engineers, keystone species, and invasive species within landscapes; and the concept of ecological disturbance.

Landforms and Biogeomorphology

In Chapter 8, a subset of the basic processes of geomorphology (aeolian, fluvial, mass movement, and coastal) was presented in the context of their roles as transport vectors for energy, materials, and information in landscapes. Weathering and glacial movement were acknowledged to be of great importance in shaping landscapes but of limited significance as vectors operating within the temporal scale relevant to landscape ecology. The landscapes we observe, study, and manipulate are founded on landforms resulting from the operation of the geomorphic processes, i.e., landforms are a consequence of entity transport in the landscape environment. Obvious examples of landforms include mountains, plains, plateaus, alluvial fans, fluvial fans, playas, dunes, barrier islands, etc. Recall, that a landform was defined to include surface geometry and underlying geologic material. Surface features are most often used to distinguish basic elements of the landscapes. However, the underlying geologic material, on which the surface geometry is displayed, may not be evident because of vegetation cover and various types of landscape-use practices. A good place to begin any landscape ecological investigation is at the landform (Phillips 1999, Stallins 2006).

Including landforms in the discussion of landscape ecology draws attention to the importance of underlying geologic material in shaping the surface geometry of the landscape. Furthermore, the means of landform development and change, as a consequence of the operation of the geomorphic processes, provide insight into how the entities (energy, information, and materials) move within and between landscapes. Together, the basic elements of landscape structure (patches, corridors, the matrix, and ecotones) form the tapestry that drapes the landform.

Several specific landscape ecological effects of landforms have been identified (Swanson et al. 1988). Landforms produce environmental gradients as a consequence of variations in slope, height, aspect, exposure, etc. Gradients can affect the distribution and abundance of living organisms. For example, plant species abundance and diversity on a mountain landform are greatly influenced by gradients resulting from variations in sunlight exposure, elevation, temperature, water retention, and soil quality. The values of these variables differ from mountain top to bottom and by cardinal direction. Landforms influence movement of the entities through effects created by gravity (mass movement), by directing wind currents (aeolian transport), and through channeling water movement (fluvial transport). In effect, the landform provides corridors for movement. The movement can be enhanced (the conduit function) or impeded (the barrier function). Landforms influence the propagation of disturbance effects in landscapes (discussed below). An obvious effect of landforms in mediating disturbance is the role that river channels and floodplains play in containing flood waters. Landforms, the geomorphic processes, and climate interact and thereby influence landscape function, i.e., the flux of energy, information, and materials in the landscape environment.

However, the landforms resulting from the earth surface processes are also influenced by living organisms. This combined view of geomorphology and ecology is referred to in the literature as *biogeomorphology* (Viles 1988). It is a concept that emphasizes the two-way interplay between ecological and geomorphological processes. The basic premise of this interaction is that the distribution and abundance of species is often related to the underlying geomorphological landform while surface morphology may in turn be altered by living organisms (Naylor 2005). This dichotomous view of biophysical interaction provides challenging research questions for landscape ecologists, e.g., What are the influences of landforms/geomorphology on the distribution and abundance of living organisms, and/or what are the influences of living organisms on earth surface processes and the development of landforms (Viles 1988)?

Investigations of biotic and geomorphological interactions are often centered on three basic processes: bioerosion, bioprotection, and bioconstruction. *Bioerosion* deals with weathering and/or erosion of the land surface by organic means. In terrestrial landscapes and marine environments, this process generally follows from boring, drilling, and burrowing activities of living organisms (Figure 9.2). *Bioprotection* deals with the roles of organisms in preventing or reducing the action or impact of other earth surface processes, e.g., the role of vegetation in controlling island bar development in fluvial systems (Figure 9.3). *Bioconstruction* deals with the production of sedimentary deposits, accretions, or accumulations by organic means, e.g., reef development (Naylor et al. 2002, Naylor 2005) (Figure 9.4). In the following section we examine the activities of organisms involved in biogeomorphology and landscape ecology in the context of ecosystem engineers (Jones et al. 1994 and 1997) and zoogeomorphologists (Butler 1995), keystone species (Davic 2003), and invasive species (With 2002).

Figure 9.2 — Bioerosion results from weathering and/or erosion of the landscape surface by organic means. In terrestrial landscapes and marine environments, this process often follows from boring, drilling, and burrowing, activities. In this figure, bioerosion is caused by burrowing activities of feral swine (Texas AgriLife Extension Service photograph).

Figure 9.3 — Bioprotection deals with the roles of organisms in preventing or reducing the action or impact of processes, such as fluvial transport. This figure illustrates vegetation cover along the banks of a river that serves to ameliorate erosion (photograph by R. F. Billings).

190 | *Landscape Change*

Figure 9.4 — Bioconstruction deals with the production of sedimentary deposits, accretions, or accumulations by organic means. Mangroves (*Rhizophora* spp.) are found in tropical and sub-tropical marine estuaries. Once established, mangrove roots provide habitat for oysters (Ostreidae). This interaction enhances sediment deposition and leads to landform development. Mangroves also provide biopretection by preventing erosion from storms (CONABIO-SEMAR photograph).

Activities of Living Organisms

In some measure, all living organisms influence the resources and conditions of their physical habitat. The details of this type of interaction fuel the curiosity of many ecologists. In our examination of the activities of living organisms and their effects on landscape-cover change, we employ a functional approach. In particular, we consider living organisms as ecosystem engineers, as keystone species, and as invasive species. Our focus is on how the organisms, classified in this manner, influence landscape-cover change.

Living Organisms as Ecosystem Engineers within Landscapes

Physical ecosystem engineering is a metaphor that describes a type of functional interaction between living organisms and their abiotic environment that results in habitat change. The concept of physical ecosystem engineers was introduced by Jones et al. (1994 and 1997), and the subject has been the center of considerable commentary since then (Wright and Jones 2006). For introductory purposes, *physical ecosystem engineering* is defined as change to the abiotic environment caused by the activities of living organisms that results in the creation, modification, maintenance, or destruction of habitats, i.e., structurally mediated modification of habitats by organisms. *Ecosystem engineers* are the organisms that mediate the changes to the abiotic environment through their influence on the resources and conditions that define habitats for the community of life that forms the biotic component of the ecosystem. An essential constraint to the concept of ecosystem engineers and engineering is that direct provisioning of resources to other species, in the form of living or dead tissues, is not engineering. The engineer has to act in a capacity beyond being the resource (Jones et al. 1994). Below, we expand on these definitions to examine the elements of ecosystem engineering important to landscape ecology in general and landscape-cover change in particular.

Ecosystem engineering and landscapes — The relevance of ecosystem engineering in landscape ecology centers on the concept when viewed in a spatial context. In this perspective the *ecosystem* is defined to be a spatially explicit unit consisting of abiotic and biotic components integrated by energy flow and nutrient cycling (Figure 4.5). An ensemble of spatially juxtaposed ecosystems (ecotopes) forms the *landscape mosaic* (Figure 7.43) (Chapters 5 and 7). The outcome of ecosystem engineering is interpreted in terms of impact on habitats of living organisms occurring within the specific ecosystems. Herein, we have previously defined *habitat* as the physical place where an organism either actually or potentially lives. This place provides both resources and conditions needed by the organism for survival, growth, and reproduction. Recall that *resources* are things consumed by an organism, i.e., quantities that can be reduced by the activity of living organisms (energy, materials, habitat). *Conditions* represent the state of the environment as defined by variables such temperature, humidity, wind speed, etc. In essence, ecosystem engineering is the creation, destruction, or modifications of habitats (Crooks 2002). The activities of living organisms, acting as ecosystem engineers, can affect directly and indirectly both the resources associated with and the conditional state of the habitat. When the effects of the organisms impact or influence more than one of the ecosystems in a mosaic, engineering becomes a landscape phenomenon.

The strategies of ecosystem engineering — How do living organisms acting as "engineers" affect ecosystems and the landscape environment? In the physical modification of habitats, living organisms employ two fundamentally different strategies: autogenic and allogenic engineering. Each approach is briefly described below.

Autogenic engineers are organisms that change the landscape environment as a consequence of their inherent physical structures, i.e., their living or dead tissues. The organisms themselves are part of the engineered habitats. The most often cited examples are forest trees and coral reefs. The development of the forest or reef results in physical structures that change the conditional state of the environment and influence the distribution and abundance of resources. For example, attributes of forests such as their species composition, age-class distribution, density, and size ameliorate the conditional state of the environment through effects on variables such as light penetration, temperature, humidity, wind speed and direction, water retention and movement, etc. The forest structure also provides the resources (energy, materials, and habitat) needed by the organisms living there (Figure 9.5) (Holling 1992).

Allogenic engineers change the landscape environment by transforming living or non-living materials from one physical state to another, via mechanical or other means (Jones et al. 1994). The "engineer" is not necessarily a part of the permanent physical ecosystem structure that results from its labors. Quintessential examples of allogenic engineers are beavers and earthworms.

In the case of beavers (*Castor canadensis*) engineering involves dam building. Wood and brush, gnawed down by the beaver, and mud are the primary materials used in construction (Butler 1995). This engineered structure has manifold effects on the stream as well as adjacent terrestrial ecosystems. In addition to changing stream morphology and hydrology above and below the dam site, conditions and resources of habitats for a myriad of living organisms are modified. In some instances habitat is lost and species abundance and diversity are reduced, and in other cases they are increased, and living organisms flourish (Naiman 1988). Eliminating beavers has a dramatic cascading effect on many taxa as well as the basic ecosystem processes (Chapter 4). Furthermore there is no

Figure 9.5 — Ecosystem engineers are organisms that mediate changes to the abiotic environment through their influences on the resources and conditions that define habitats for living organisms. Autogenic engineers are organisms that change the landscape environment as a consequence of their inherent physical structure, e.g., a forest (photograph by R. F. Billings).

redundancy built into the roles that beavers play in landscapes (Lawton 1994). Eventually sediment infilling occurs because beaver dams reduce the ability of a stream to transport sediment by lowering the effective slope of the channel. The side effect of this infilling is significantly cleaner water below the dam (Butler 1995) (Figure 9.6).

In the case of earthworms (Lumbriculidae), engineering occurs in two ways. The direct effects of engineering take place as a result of cast production. This process results in microlandform creation (Butler 1995). Earthworm castes consist of mixed organic and inorganic materials from the soil that have been voided after passing through the intestine. Surface casting is an essential process that creates habitat in earthworm communities (Figure 9.7). The indirect effects of burrowing, mixing, and casting activities of earthworms influence the processes of infiltration, soil creep, surface wash, and rainsplash detachment (Butler 1995). In turn, these processes affect nutrient cycling, hydrological flow, rates of soil erosion, etc.

Living Organisms as Keystone Species in Landscapes

Keystone species represent a second type of functional group of organisms that can be important agents of landscape-cover change. A *keystone species*

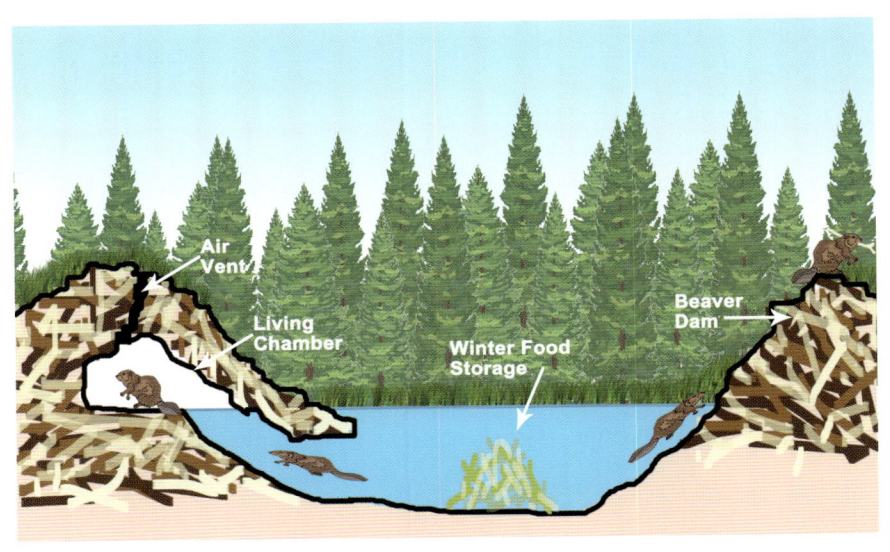

Figure 9.6 — Allogenic ecosystem engineers change the landscape environment by transforming living or non-living materials from one physical state to another via mechanical or other means. The American beaver (*Castor canadensis*) is an example of an allogenic engineer (KEL drawing).

Figure 9.7 — The microlandform creation resulting from the burrowing, mixing, and casting activities of earthworms (Lumbriculidae) is an example of allogenic engineering (Wikimedia Commons photograph).

is one whose effect on an ecosystem is disproportionately large relative to its low biomass in the community as a whole (Power et al. 1996). The term *keystone* is a metaphor that relates the importance of selected species in an ecosystem with the central wedge in an arch of stones that locks the parts together. Keystone species are considered to function as the central supporting element in the biotic community of the ecosystem. As with our discussion of ecosystem engineers, the topic of keystone species has been the subject of considerable commentary by ecologists (Power et al. 1996, Davic 2003) since it was introduced and defined by Paine (1966, 1969).

In contrast to the mode of action of ecosystem engineers, which involved structurally mediated modification of habitats, keystone species influence the ecosystem components of landscapes as a consequence of trophic relations. Among ecologists, a great deal of interest in the community of life associated with ecosystems is directed to interactions among species, which is often represented by a food web that defines the feeding relations present. The keystone species concept centers on the proposition that the functional attributes of the species forming the biotic community of the ecosystem are as important as the number of species present (Hooper et al. 2005). In a community, keystone species are more important to ecosystem functioning than would be suggested simply by biomass. For example, the tsetse fly (*Glossina* spp.) in central Africa occurs in low biomass. However, this insect has a significant community impact as it vectors a disease that is lethal to a variety of mammal species, including humans. The disease, sleeping sickness, is caused by a parasitic flagellate protozoan (*Trypanosoma* spp.), and it eliminates certain mammal species from savanna, riverine, and forest ecosystems. This insect has a large effect on ecosystem processes per unit of tsetse fly biomass, because it limits the density and distribution of many ecologically important mammals (Figure 9.8) (Chapin et al. 2002).

Following from the broad-based definition of keystone species (presented above) and the emphasis on functional roles and trophic interactions of living organisms in ecosystems, a variety of species can be afforded the status of "keystone." These organisms have been variously labeled as keystone herbivores (bark beetles in pine forests), predators (coyotes in deserts), prey (stoneflies in trout streams), pollinators (honey bees in agricultural fields), hosts (fig trees in tropical forests), etc. Loss of a keystone species from a community has a greater ecological impact than does the disappearance of a species from an ecologically similar functional group. For example, many species of bark beetles infest southern yellow pines, *Pinus* spp., but only herbivory by the southern pine beetle, *D. frontalis,* is capable of altering forest landscape structure to an appreciable extent. The landscape ecological tie to this concept of keystone species centers on how living organisms involved in the ecosystem processes of primary production, consumption, and decomposition (Figure 4.5) create spatially explicit changes in landscape-cover through their interactions and activities. The discussion of ecosystem engineers emphasizes physical modification of the landscape environment while the treatment of keystone species emphasizes trophic interaction, i.e., feeding relations as defined through food web interaction. The consequences of the feeding relations can directly and indirectly cause landscape-cover change.

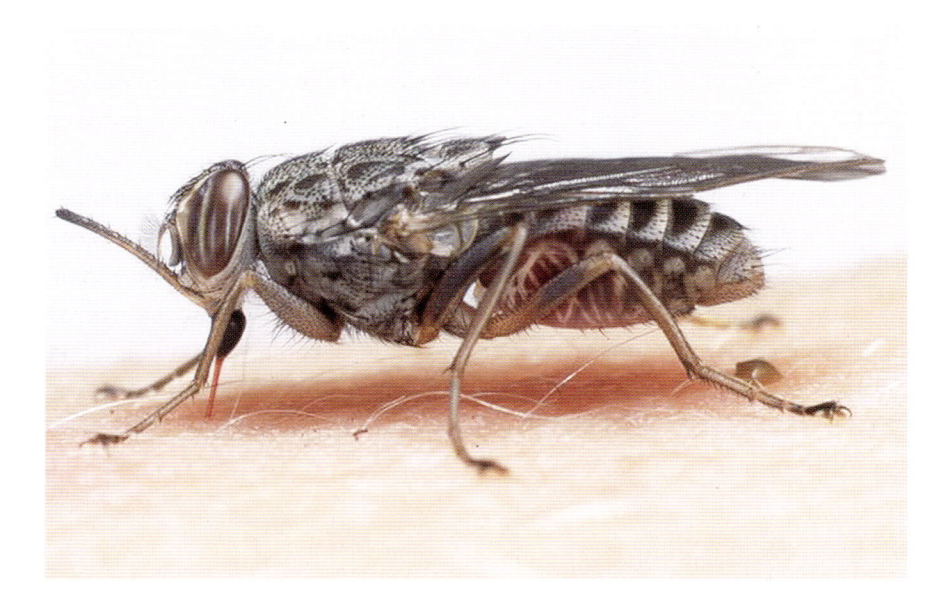

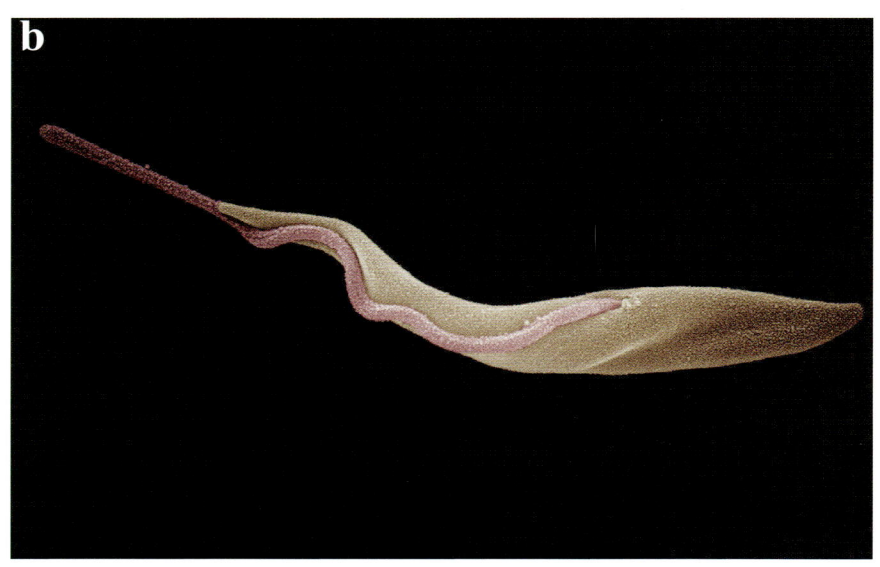

Figure 9.8 — The tsetse fly, *Glossina* spp. (Diptera: Glossinidae) is a keystone species, i.e., a species whose effect on an ecosystem is disproportionately large relative to its low biomass in the community as a whole. Tsetse flies occur in central Africa in low biomass (a). However, this insect has a significant community impact as it vectors a disease (sleeping sickness) that is lethal to a variety of mammal species, including humans. Sleeping sickness is caused by a parasitic flagellate protozoan (*Trypanosoma* spp.) (b) and it eliminates certain mammal species from savanna, riverine, and forest ecosystems (a, Department for International Developement-Animal Health Programme photograph, b, Wikimedia Commons photograph).

Living Organisms as Invasive Species in Landscapes

The discussion of ecosystem engineers and keystone species emphasized the functional roles of organisms and their affects on landscape-cover change. Invasive species represent a third categorization of living organisms that are important to our discussion of landscape change. Although there is a significant ecological underpinning to the subject of invasive species, the genesis of interest in these organisms and their associated behavior is clearly anthropocentric. The scientific and social agenda of invasive species is broad-based and includes topics such as evaluating and managing ecological and economic impact of invaders; restoration of landscape resistance to invasion; distribution mapping; spatial epidemiology; community interactions among invasive and indigenous species; monitoring movement and introduction; effects of landscape structure on movement, distribution, and escape of transgenes; etc. (Stohlgren and Schnase 2006, Lockwood et al. 2007, Buckley 2008). In this section we examine two aspects of invasive species in the context of landscapes: invasive species and landscape-cover change and the effects of landscape structure on the invasion process.

Invasive species and landscape-cover change — Various terms are used to describe species which have been introduced into novel ecosystems: e.g., alien, exotic, non-indigenous, non-native, invasive, etc. *Invasive species* is the most common descriptor, and the organisms classified in this manner are the subject of considerable scientific inquiry (Sakai et al. 2001, With 2002, Didham et al. 2005, MacDougall and Turkington 2005), regulatory scrutiny (Stohlgren and Schnase 2006), and social/political commentary (Buckley 2008).

In the context of landscape ecology, an *invasive species* is a non-indigenous (non-native) organism that has been introduced into a novel ecosystem, i.e., an ecosystem within a landscape that occurs outside the natural range or potential dispersal distance of the species. The anthropocentric definition of invasive species is as follows: "an alien species whose introduction does or is likely to cause economic or environmental harm or harm to human health" (National Invasive Species Management Plan). In this context, for a non-indigenous organism to be considered an invasive species, the negative effects that the organism causes, or is likely to cause, are judged to outweigh any beneficial effects. Clearly, the term *invasive species* is a metaphor that relates the introduction of non-indigenous species to military conquests of new territories. The attendant vocabulary carries this emphasis as well: invaders, alien, barriers to entry, lines of defense, devastating impact, threats to biodiversity, biosecurity, harm, etc. These terms are certainly descriptive and relate the fact that there is public interest in and perhaps awareness of the consequence of introducing non-indigenous organisms into novel ecosystems. In the United States, the National Invasive Species Information Center, in the Department of Agriculture (USDA), serves as a clearing house for information from Federal, State, local, and international sources <http://www.invasivespeciesinfo.gov>.

Non-indigenous species have been distributed throughout the world. In some instances the introductions have been intentional and beneficial. In other instances the introductions have been unintentional with significant negative ecological and economic impact. In still other instances the introductions have been purposeful but unadvised. Each of these circumstances is examined below.

Many of the introductions of invasive species into novel ecosystems have been intentional from a human perspective, e.g., domesticated food plant and animal species. In general, species under domestication, cultivation, or human control (horticultural species, pets, etc.) are not considered to be invasive species, as the beneficial effects derived from these organisms are judged to outweigh negative impacts. Knowing and purposeful introductions are by far the most common. However, the landscape ecological consequences of cultivation and husbandry are often immense, e.g., replacing the short-grass prairie landscapes of the Great Plains with agricultural crops and substituting bison with cattle. Although recovery efforts from unadvised introductions are a significant human enterprise, much of the emphasis in invasive species centers on prevention of new introduction and evaluating the potential consequences of introductions once they do occur.

In other instances, the introductions have been unintentional with significant negative landscape ecological as well as social impacts. For example, the inadvertent (accidental) introduction of the brown tree snake (*Boiga irregularis*) (Figure 9.9) in Guam in the 1950s has had a significant impact on native forest vertebrate populations and human health. In particular, populations of birds, bats, and reptiles were depredated; and by 1990 most forested areas of Guam retained only three native vertebrate species (Fritts and Rodda 1998). Brown tree snakes also are responsible for considerable social/economic impact resulting from costs associated with medical treatment of snakebites, snake-caused power outages ("brown outs"), and decreased tourism. Furthermore, as Guam is a major transportation hub in the Pacific, prevention activities associated with exporting this organism require constant vigilance with attendant expense. The potential impact of the introduction of this organism to other Pacific islands is significant (Shwiff et al. 2010) and an issue of considerable concern.

In still other instances, introductions of invasive species have been intentional but resulted in unanticipated as well as negative ecological and economic impact. Classic examples include kudzu and nutria. Each case is examined below.

Figure 9.9 — The brown tree snake (*Boiga irregularis*) is an invasive species that was introduced into Guam in the 1950s. This species has had a significant impact on native forest vertebrate populations and human health. In particular, populations of birds, bats, and reptiles were depredated and by 1990 most forested areas of Guam retained only three native vertebrate species (Wikimedia photograph).

Kudzu, *Pueraria montana*, is a woody vine purposefully introduced into the southern United States from Japan and China in the 1900s (Miller 2003). The intended purpose was to mitigate soil erosion associated with abandoned cotton farms. Although persistent, the root system of kudzu did not protect soils from erosion; and the vegetation growth of the vines covered native plants, ground surfaces, and structures (Figure 9.10). Ironically, one approach being pursued to manage the negative impacts of kudzu is to identify and introduce an herbivorous biological control agent, an approach successfully used to manage prickly pear cactus (*Opuntia* spp.) in Australia.

The nutria, *Myocastor coypus*, is an herbivorous rodent species native to South America that was purposely introduced along Gulf Coast of the United States, perhaps as early as the 1890s (Figure 9.11). The native habitat for this species is coastal lakes and marshes, and it was well-suited to the new landscape environment. The initial intent of the introduction was to provide a new species for the fur trade. When the economic market for nutria fur became unprofitable in the 1940s, this species was released from captivity and is now established in 16 states, ranging from the Gulf Coast to as far north as the Delmarva marshes of Chesapeake Bay in Maryland. The negative impacts of nutria are manifold. The principal landscape ecological consequence involves damage to coastal wetlands through consumption of native vegetation. This activity results in soil erosion that affects stability of the marsh landscape, plant community species composition, and associated wildlife species. The burying activities of nutria weaken flood control levees and dam enclosures constructed for rice cultivation. This invasive species also consumes human food crops (rice, soybeans, and sugarcane) and transmits diseases.

Landscape structure and invasive species — Because of the worldwide ecological, economic, and social impact of invasive species, there is considerable scientific and political interest directed to regulation, monitoring, and managing the introduction and spread of new non-indigenous organisms into novel ecosystems. The starting point for understanding the interaction of invasive species and landscape-cover change is a definition of the general process of species invasion. The process is envisioned to consist of five sequential stages: introduction, colonization,

Figure 9.10 — Kudzu, *Pueraria montana*, is an invasive species purposefully introduced into the southern United States from Japan and China in the 1900s. The intended purpose was to mitigate soil erosion associated with abandoned cotton farms. Although persistent, the root system of kudzu did not protect soils from erosion; and the vegetation growth of the vines covers native plants, ground surfaces, and structures (photograph by Galen P. Smith/Wikimedia Commons).

Figure 9.11 — The nutria, *Myocastor coypus*, is an invasive species that was purposely introduced along Gulf Coast of the United States from South America, perhaps as early as the 1890s. Herbivory and burrowing activities by this rodent species damages coastal marches (photograph by Gwen Juarez/ Wikimedia Commons).

Landscape Change | 203

establishment, dispersal, and spatial distribution of populations (With 2002) (Figure 9.12). Each stage in the invasion process is consequential in evaluating the likelihood that a population of a non-indigenous species will survive, grow, and reproduce if introduced into a new landscape environment. If the species does persist it will influence landscape-cover change through activities associated with ecosystem engineering or trophic interactions.

As with ecosystem engineers and keystone species, invasive species become a landscape ecological issue when their presence is considered in a spatially explicit manner. Following we examine invasive species populations in the context of the steps in the invasion process. Emphasis placed on how landscape structure influences invasive species at each step in the process. However, the interaction is reciprocal, i.e., the presence of invasive species also results in landscape-cover change.

The introduction of an invasive species (step 1) into a novel ecosystem within a landscape is a function of the operation of the basic transport vectors (Chapter 8). Human transport, whether intentional or accidental, is by far the most common means of introduction, and

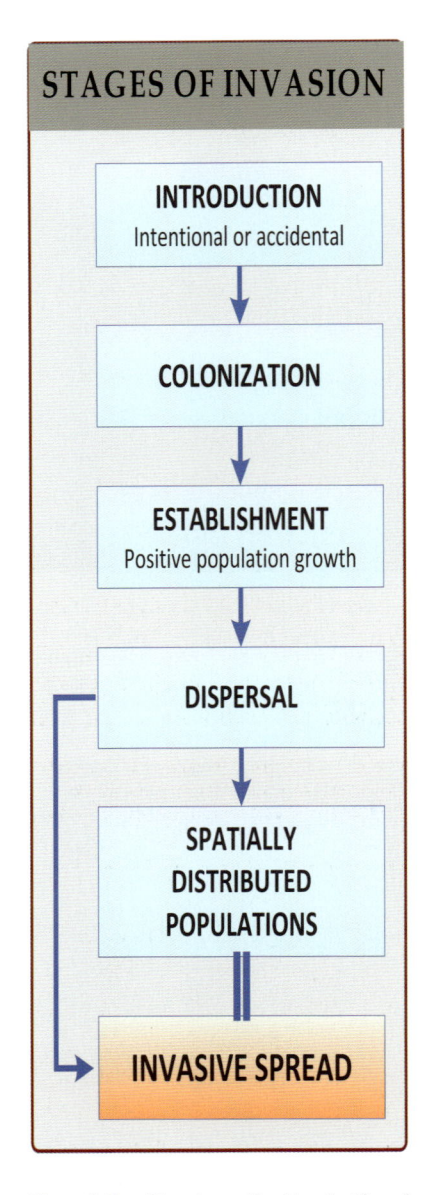

Figure 9.12 — There is considerable scientific and political interest directed to regulating, monitoring, and managing the introduction and spread of new non-indigenous organisms into novel ecosystems. The process of invasion is envisioned to consists of five basic steps: introduction, colonization, establishment, dispersal, and spatial distribution (modified from With 2002).

landscape structure is not likely to play a major role in the initial arrival of an invasive species, given this circumstance. Human transport circumvents natural geographic barriers to movement. When a non-indigenous species is purposefully introduced, for example when a biological control agent is released to "combat" a pest species, there is a tacit assumption that the suitability of the landscape environment was assessed *a priori*.

Colonization (step 2) involves the initial interaction of an invasive species with the new landscape environment. To some degree serendipity and persistence play a role in whether or not the invasive species can survive in the new environment. Successful colonization may involve repeated introductions to overcome obstacles such as the discovery of suitable spatially distributed and perhaps ephemeral habitat sites, assembly of a population size large enough to allow for reproduction, seasonal timing of environmental events relative to the natural history of the organism, etc. The opportunity for trial and error colonization events resulting from exposure of non-indigenous species into novel landscapes has greatly increased as a consequence of the immense transportation network of wolrdwide commerce.

The establishment (step 3) of an invasive species means that the organism has discovered a landscape environment where it can survive, grow, and reproduce. This environment can consist of a single ecosystem or an *ecosystem cluster* that provides requisite conditions and resources needed by the organism. The establishment stage involves *maintenance movement* within and among connected ecosystems or clusters. Recall that the motivational states of the species guide when movement takes place, where it leads, and how far it goes. The affects of landscape structure on movement are governed by the general rules identified in Chapter 7.

Once an invasive species becomes established, persistence requires location and colonization of additional suitable habitat sites, distributed within the landscape (step 4). Colonization of these new sites involves vector-mediated dispersal, i.e., movement of an individual or population away from its place of birth or origin, which is influenced by the composition and arrangement of the constituent ecosystems forming the landscape. *Functional heterogeneity* (Chapter 7), in reference to a specific invasive organism, will establish the degree of connectivity of the landscape. Landscapes where suitable habitat sites are numerous and connected are easily colonized and exploited by invasive species. By contrast, long dispersal distances and patchy habitat distribution inhibit colonization and exploitation. The various costs and benefits of dispersal were discussed in Chapter 7.

Spatially distributed populations of invasive species within a landscape represent the endpoint of the invasion processes (step 5). The establishment of spatially distributed self-sustaining populations (metapopulations), within the shifting mosaic of ecosystems that form a landscape, enhances the likelihood that the species will persist in the presences of various types change events. The consequences of establishment of spatially distributed populations of an invasive species represent a focal point of interest among landscape ecologists. Three issues are of noteworthy concern: competitive interactions among native and non-indigenous species, response of non-indigenous species to natural and cultural disturbance, and role of non-indigenous species in the adaptive cycle of ecosystem change. Each of these subjects is discussed below.

First, one of the key attributes associated with many invasive species is that that they have a much greater impact on the new landscape than they did in their native environment. An obvious explanation is that the new landscape environment is void of natural enemies, and this circumstance affects demographics through increased natality and reduced mortality of the invasive species (Pimentel 1963), i.e., competitive interactions favor the non-indigenous organism. Another common observation is that the interactions between native and non-indigenous species are initially virulent and that with time and adaptation the association becomes less severe, i.e., attenuated. Examples of particularly virulent interactions of invasive species in forest landscapes in the United States include the emerald ash borer (*Agrilus planipennis*) (Figure 5.14), which is lethal to several species of ash (*Fraxinus* spp.), and the hemlock woolly adelgid (*Adelges tsugae*), which is a mortality agent of Eastern hemlock (*Tsuga canadensis*) and Carolina hemlock (*T. caroliniana*) (Figure 4.6a). In both cases the effect on landscape-cover is the elimination of dominant host trees species from the landscape environment.

Second, invasive species are frequently associated with natural and cultural disturbance to landscapes (Hobbs and Huenneke 1992). Disturbances (discussed in the following section) commonly result in habitat loss and an accompanying decline in species diversity. A frequent observation is that there is a positive correlation between native species decline and invasive species dominance in disturbed landscapes. However, it is not clear that the invasive species directly cause the decline through competitive superiority. An alternative explanation is that the invasive species take opportunistic advantage of disturbance effects, such as habitat loss, and are not actually drivers of the change (MacDougall and Turkington 2005).

Third, invasive species introduction represents a serendipitous event that can trigger adaptive ecosystem change (Figure 4.7). The transition from the back loop to the forward loop ($\alpha \rightarrow r$) in the adaptive cycle is where the change in the trajectory of ecosystem development can take place. In effect, the activities of the invasive species represent an experiment that can result in a restructuring or destruction of the ecosystem. The change in ecosystem development is propagated into the landscapes through the panarchy.

Ecological Disturbance

Disturbances are initiating causes that often result in landscape-cover change. Recall that an *initiating cause* was defined simply to be a factor that brings about an effect or a result, e.g., a lightning strike igniting a tree in a forest (Chapter 8). Our treatment of the topic of disturbance follows from the previous discussion of the cause and consequence relation summarized in the general transport model, as illustrated in Figure 8.2. In effect, the various types of natural and cultural disturbances represent flavors of initiating causes. In this section three aspects of disturbance ecology are examined: the definition and use of the term *disturbance* as applied to landscapes, the impact of disturbances on landscapes, and the concept disturbance regime.

Disturbance in Landscape Ecology

Although the term *disturbance* is used commonly in daily parlance and it is a fixture in the ecosystem and landscape ecology literature, there is no generally accepted and unambiguous definition[1]. Again, the concept of disturbance is anthropocentric and therefore open to subjective and specialized interpretation. A further complication is that the terms *perturbation* and *stress* are frequently used as synonyms for disturbance. Environmental scientists have extended the vocabulary by adding the word *stressor* (something that causes stress) to the mix.

Because the term disturbance is used in a variety of ways, a precise definition that accommodates all applications is not necessary or possible. For our purposes, we tailor a definition of disturbance to accommodate the previous discussion of landscape function (Chapter 8). The definition specifies the circumstances under which an initiating cause/consequence relationship is considered to be a disturbance, in contrast to a normal or expected event. Our working definition of *disturbance* is as follows: an initiating cause (physical force, a process, or an event) that produces an effect (consequence) that is greater than average, normal, or expected. This definition requires a reference state, i.e., a mean condition bounded by a range in variation. In some instances this normal or average state can be easily defined based on historical records or experiential knowledge. For example, weather station data (temperature, humidity, precipitation, wind speed, rainfall, etc.) provide a record of average atmospheric conditions. Most precipitation events at a location occur within a defined range, e.g., number of events/per year, number of cm of rainfall/hour, etc. Using our definition, if a specific precipitation event exceeded the average by two or three standard deviations, then it could be classed as a disturbance. The ecological response to or consequence of such an event would also be measurable, e.g., increased soil erosion, beyond that expected for a normal precipitation event. The classification scheme used to characterize initiating causes (Chapter 8) can be directly applied to disturbance events as well. By contrast, in other instances, the periodicity and severity of a disturbance event is unknown or the data describing it are inadequate to

---

[1] A frequently cited definition is as follows: "A disturbance is a relatively discrete event in time that disrupts ecosystem, community, or population structure and changes resources, substrate availability, or the physical environment" (Pickett and White 1985). What does "relatively discrete" mean to you?

define a nominal condition. In this circumstance disturbance designation is the subject of human judgment.

Defining a disturbance event in terms of deviations from a mean condition or state requires that we further specify time and space boundaries. The impact of most recurring natural disturbance events will average out in terms of both frequency of occurrence and intensity, if observed over a long enough period of time, e.g., 10,000 years. This same generality pertains to the spatial extent of disturbance events. For example, under most circumstances, wildfires and bark beetle herbivory are both considered to be disturbances in forest landscapes. When these events occur, substantial effort is often directed to reducing their impact by application of remedial treatment tactics, i.e., fire control and insect suppression. However, Cairns et al. (2008) and Xi et al. (2009b) demonstrated, through simulation modeling experiments covering a period of 500 years, how landscape structure, forest fires, and bark beetle herbivory interact to produce sustainable pine forest landscapes in the southern Appalachian Mountains of the United States. Therefore, a practical time and space constraint needs to be placed on evaluation of disturbances effects. A time frame of approximately five human generations (Forman 1995) provides a practical boundary. Most genetics studies use a period of 20 to 25 years for a human generation, but the number varies with culture, gender, and other circumstance. The spatial framework for evaluating disturbance events, again, can be specified for practical purposes to include the landscape of interest, although the impact may extend over a much greater extent.

Impact of Disturbances on Landscapes

Herein the term *impact* is defined to mean any effect on the landscape environment resulting from a disturbance event. Disturbances affect landscapes in multiple ways and can be evaluated from, social, economic, political, and ecological perspectives. *Social* (axiological) impact refers to the effects of disturbance on aesthetic, moral, and metaphysical values associated with the landscape, e.g., fire scars altering the appearance of a scenic vista. *Political* impact refers to the effects of disturbances on the landscape environment that result in actions, practices, policies of local, state, or federal governmental agencies, e.g., laws to regulate pollution of streams. *Economic impact* is simply defined as the effect of disturbances on the monetary receipts from the production of goods and services from a landscape, e.g., loss of revenue resulting from an insect or disease outbreak that killed merchantable trees. Given our concern for the biophysical attributes of the landscape environment, i.e., landscape-cover change, we are particularly interested in ecological impacts. *Ecological impact* is the qualitative or quantitative change in conditions and/or resources in the landscape environment resulting from a disturbance event. As with initiating causes, disturbance events vary in space and time and can be biotic or abiotic, exogenous or endogenous, natural or anthropogenic, and discrete or chronic (Coulson and Stephen 2006).

The scope of the concept of ecological impact is immense, however, there are several recurrent themes which include the following: effects of disturbances on landscape transformation, primary production, nutrient cycling, biodiversity, endangered and threatened species, and population dynamic of selected species. Each of these themes is briefly examined below.

Disturbance Effects and Landscape Transformation — *Transformation* deals with change in form, appearance, nature, or character of the landscape. Disturbance can create different types of structural transformations in landscape through perforation (e.g., forest fires), dissection (e.g., construction of road corridors), fragmentation (e.g., urbanization), shrinkage (e.g., saltwater intrusion into wetlands), and attrition (e.g., conversion of forestland to farmland) (Forman 1995). Transformation is discussed in more detail in the section of this chapter dealing with pattern formation in landscapes.

Disturbance Effects on Primary Productivity — Landscape transformations, in turn, affect the basic ecosystem processes. Recall from our discussion of ecosystem function that *primary productivity* is the rate of conversion of solar energy by green plants via photosynthesis to organic substances. Agriculture, forestry, and range management deal in large part with regulating primary production. Insect and disease outbreaks and overgrazing represent disturbance events that can have a major impact on primary production and therefore are of considerable ecological as well as economic interest. These managed landscapes are also subject atmospheric disturbances, e.g., wind, ice, and hail storms.

Disturbance Effects on Nutrient Cycling — Recall, again, from our discussion of ecosystem function that nutrient cycling deals with the transformation of chemical elements from inorganic form in the environment to organic form in organisms and, via decomposition, back into inorganic form. The landscape ecological emphasis on disturbance effects related to nutrient cycling centers on processes such as soil erosion, nonpoint source pollution, impact of atmospheric deposition, carbon sequestration, etc. Investigations of the effects of change in nutrient dynamics have been a focal point in ecosystem research for many years (Bormann and Likens 1979, Coleman and Hendrix 2000).

Disturbance Effects on Biodiversity — Habitat loss, land transformation, and the activities of invasive species are generally recognized to be the greatest threats to biodiversity. *Biodiversity* is defined as the variety and abundance of species that occur in an ecosystem or landscape of interest. For practical purposes, biodiversity is synonymous with *species diversity*. Studies of species diversity generally deal with both the number of species present (richness) and the proportion of individuals associated with each species (evenness). Change in species diversity resulting from disturbance events is an important issue in landscape ecology, because of the threat to valuable intermediate and final ecosystem services (Chapter 4) provided by living organisms, e.g., pollination of flowering plants, biological control of pest species, pure air, clean water, soil formation and protection, waste recycling, nitrogen fixation, carbon sequestration, food, fuel, fibers, medicine, ecotourism, etc.

Disturbance Effects on Rare and Endangered Species — Rare and endangered species represent a subplot in the larger story of biodiversity. In the broad view, emphasis is placed on protecting species, populations, and genetic resources. In the United States, the Endangered Species Act (ESA) of 1973 (as amended) provides for protection of endangered or threatened species of animals and plants. The substantive requirement of the ESA is that persons cannot unlawfully *take* any officially listed endangered or threatened species. The term *take* has evolved to mean not only killing the species, but also destroying critical habitat. Therefore, virtually any change in the landscape environment could potentially affect a suite of endangered or threatened species.

Disturbance Effects on Population Dynamics of Selected Species — Disturbances can affect population dynamics of species that are identified to be of particular importance. In general, an organism is singled out for special treatment because it is an ecosystem engineer, a keystone, an invasive, an endangered, or a pest species. However, in some instances organisms are valued because of their importance as game species (e.g., the wild turkey, *Meleagris gallopavo*) or because it is a rare or interesting species prized by ecotourists (e.g., the grizzly bear, *Ursus arctos horribilis*). In some instances, disturbances may reduce or degrade habitat needed by a species, and thereby result in a decline in population size. In other instances, disturbances in landscapes create favorable habitat, which results in an increase in population size. This circumstance is often associated with pest species outbreaks. A poignant example of a natural disturbance affecting population abundance of a species was the impact that Hurricane Hugo had on the Red-cockaded woodpecker (*P. borealis*). In 1989, Hurricane Hugo passed through the Francis Marion National Forest in South Carolina. Wind damage from this category 5 hurricane devastated the second-largest population of Red-cockaded woodpeckers then in existence, i.e., approximately 65 percent of the 477 known groups of birds were dead or missing after the event (Hoyle 2008).

Disturbance Regimes

During the course of a five human generation time frame, which we specified as a practical temporal boundary for evaluation and planning purposes, a given landscape will often be exposed to a variety of natural and cultural disturbance events. Taken together the various types of disturbances are referred to as a disturbance regime. In particular, a *disturbance regime* is defined as the ensemble of disturbance types associated with a specific landscape environment.

As with initiating causes, disturbance effects are spatially explicit, but may include multiple ecosystems within a landscape or even landscapes associated with different ecoregions, e.g., the storm track for Hurricane Katrina (Figure 9.13). Individual disturbance events serve as triggers for creative destruction in the adaptive cycle of ecosystem change, i.e., the transition from the forward to the back loop of the adaptive cycle (K$\rightarrow\Omega$) (Figure 4.7). Disturbance events affect the various levels in the panarchy

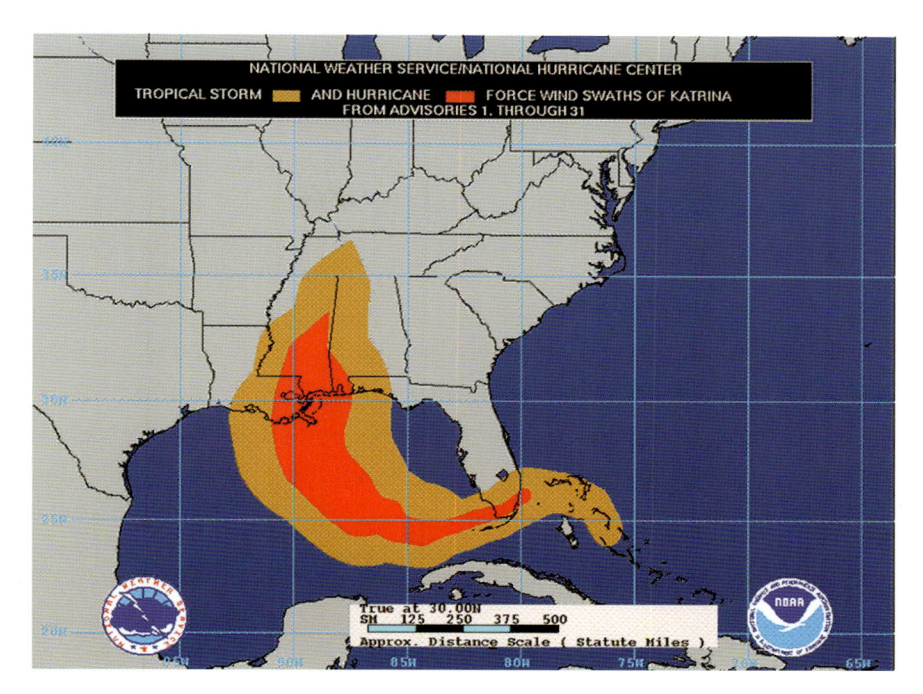

Figure 9.13 — Distrubance effects are spatially explicit, but may include multiple ecosystems within a landscape or even landscapes associated with different ecoregions. The storm track for Hurricane Katrina, which ocurred in 2005, impacted ecoregions associated with four states in the southern United States (NOAA image).

in different ways, and the ecological impact will vary as a function of spatial and temporal scale.

The application of the concept of disturbance regimes to evaluate landscape-cover change has practical implications for management and conservation. The task is to identify and sequence the various types of disturbance events associated with a specific landscape over a five-generation time span. Interpretation involves ecological impact assessment in the context of deviation from the expected trajectory of landscape change. Although this statement of task is simple, addressing it from a landscape perspective has proven to be a difficult undertaking for several reasons. First, a great deal of emphasis in ecology has been placed on the concept of ecosystem succession (Chapter 4), but this same degree of concern has not been applied to landscape change. The contemporary innovations in landscape succession are in large part associated with simulation modeling of forests

(Mladenoff 2004). The LANDIS model has been particularly useful in this regard, as it has a component that deals specifically with disturbances. Second, disturbances are usually episodic events associated with the process of landscape succession. Recall that an *episode* is an event that is distinctive and separate although part of a larger series, i.e., a situation that is integral to but conceptually separable from a continuous process or stream of events. For example, Rykiel et al. (1988) demonstrated how pine forest aging, favorable weather conditions for insect development, and lightning strike frequency and distribution interact and result in widespread outbreaks of the southern pine beetle (*D. frontalis*). Third, empirical data on natural variability associated with disturbance events is generally unknown (Landres et al. 1999). Recall, again, that our definition of disturbance was based on impact measured in terms of deviation from a nominal state. Fourth, to some degree, the impact of disturbances is shaped by past modifications of the landscapes (Peterson 2002). For example, a forest fire event will add age-class heterogeneity to the landscape and reduce the fuel load in burned patches. The impact of subsequent fires will affect the previously burned patches less than those that escaped the first event. In assessing impact, the previous disturbance history of a landscape is rarely known, but can in some cases be reconstructed, e.g., tree ring analysis to identify scars left from previous forest fire events (Figure 9.14), pollen in lake deposits, remnant seed beds, etc. In some cases, the legacies of landscape-use can be used to evaluate probable impacts of disturbance events. Fifth, component ecosystems of landscapes are often dependent on and adapted to disturbance events. For example, regeneration of the chamise-chaparral (*Adenostoma fasciculatum*) forests of California is dependent on fire. As these forests age, there is a high proportion of dead wood, little annual growth, and no new seeding development. At about 60 years, plant community development is truncated, and fire serves to reinitiate the adaptive cycle (Figure 9.15) (Hanes 1971).

Figure 9.14 — In assessing disturbance impact, the previous history of a landscape is rarely known, but can in some cases be reconstructed, e.g., tree ring analysis to identify scars left from previous forest fire events, pollen in lake deposits, remnant seed beds, etc. This photo shows eight fire scars associated with a pitch pine, *Pinus rigida*, sample collected by C. Lafon and H. Grissino-Mayer on the George Washington National Forest in Virginia (photograph by J. Wulfson).

Figure 9.15 — Regeneration of the chamise-chaparral (*Adenostoma fasciculatum*) forests of California is dependent on fire. As these forests age, there is a high proportion of dead wood, little annual growth, and no new seeding development. At about 60 years, plant community development is truncated, and fire serves to reinitiate the adaptive cycle (Chaparral Fire Research photograph).

Landscape Change | 215

LANDSCAPE-USE CHANGE

Landscape-use change deals with human purpose or intent as applied to the biophysical attributes of the landscape environment (Lambin et al. 2001). In the following sections we examine two aspects of this subject. First, we consider landscape-use change in the context of landscape domestication. Second, we examine how human needs serves as drivers of landscape-use change. Clearly, landscape-use change results in landscape-cover change, i.e., these subjects are inseparably interrelated.

Landscape Domestication and Human Needs

The human role in facilitating landscape-use change takes place as a consequence of landscape domestication practices (Kareiva et al. 2007). The term *landscape domestication* is defined to be the activities of humans that structurally shape and functionally modify landscapes to satisfy basic human needs. With some concession to simplification, the basic human needs include adequate food, water, housing, energy, health, and cultural cohesion. The behavioral sciences address the issue of human needs in a considerably more robust manner, with Maslow's hierarchy of needs often serving as a starting point (Maslow 1954). These basic human needs translate directly to landscape-use change in the form of consequential actions directed to agricultural production, natural resource management, construction and destruction of the built environment, systems of commerce, provision of public health, and cultural practices.

In the context of the transport model (Figure 8.2), actions associated with domestication are *initiating causes* that produce predictable changes in landscape-use resulting from human manipulation of energy, materials, and information within the landscape environment. In some instances the changes can be interpreted in the context of our discussion of disturbances, i.e., deviations from expected or normal conditions, e.g., deforestation of a tropical rainforest landscape to accommodate human settlement. Landscape domestication bundles the processes that lead to turning landscape space into places where the basic human needs are met. Today, landscape domestication is an intricate and worldwide phenomenon, but the genesis traces to the archeological records of early human settlement (Terrell et al. 2003).

Drivers of Landscape-Use Change

The basic human needs are the drivers of change in landscape-use. For our purposes, a *driver* is simply a human-initiated cause (physical force, a process, or an event) that results in landscape-use change. This definition is restrictive to landscape-use change as there are also drivers of landscape-cover change that are not human-initiated. So initiating causes, interaction of living organisms with the landscape environment, disturbances, and drivers are variations of the same conversation, separable by issues of emphasis. Generally, interest in drivers of landscape-use change is associated with questions dealing with location, extent, and rate (Lambin et al. 2001,Velkamp and Lambin 2001, Bürgi et al. 2004), i.e., where did the change occur, how large an area was affected, and how fast did it happen. In our discussion of landscape-cover change emphasis was placed on how landscapes change as a consequence of the interaction of landform, activities of living organisms, and disturbances. In this discussion of landscape-use change, emphasis is placed on why landscapes change. The drivers are both biophysical and anthropocentric.

Biophysical drivers

The surface geometry, landform, and climatic regime delineate logical boundaries for human purpose or intent that result in landscape-use change, i.e., these elements of the landscape environment establish where human-initiated change is likely to occur and what form it will take. Attributes of the landscape environment attract different types of domestication, e.g., fertile landforms attract agriculture and forestry, coastal landforms attract habitation and commerce, mountainous landforms attract recreationists, etc. By contrast, aeolian effects on landforms are pronounced in arid climates where vegetation cover is sparse and surface material is dry. Such landscapes would not easily lend to productive agriculture. A fluvial plain (Figure 9.16) would not be a suitable landform on which to build a habitation. Predicting where landscape-use will be changed by domestication and what form it will take is straightforward and follows from an appraisal of the suitability of the landscape for provision of human needs[2].

---

[2] How did the architects of the Great Pyramid in Egypt come to select Giza as the site for this extraordinary and massive structure, i.e., did the underlying landform have any influence?

Figure 9.16 — A flood plain would not be a suitable landform on which to build a habitation. Canterbury Plains, New Zealand (KEL image).

Anthropocentric drivers

The rate and spatial extent of landscape-use change are driven by demand for the commodities that feed the basic human needs. The term *commodity* is broadly defined to be something useful or valuable to humans. In a landscape ecological context, commodities translate directly to ecosystem goods and services (Chapter 4). The human-motivated (anthropocentric) drivers are indirect as they are actions that facilitate landscape-use change which results in landscape-cover change. The anthropocentric drivers are rooted in issues associated with demographics, economics systems, sociopolitical policy, and technological and scientific developments. A great deal of contemporary landscape-use change resulting from anthropocentric drivers is associated with deforestation, rangeland modification, agricultural intensification, and urbanization.

The deforestation of Rondônia, Brazil, provides a good example of how anthropogenic drivers operate and effect landscape-use change. Rondônia is a state in Brazil, located in the northwestern part of the country. It is a rainforest area of the southern Amazon Basin that has undergone extensive deforestation beginning in the early 1970s (Figure 9.17) and continuing into the present time. With some concession to simplification, the fundamental anthropocentric driver centers on government policy regarding landownership. The Brazilian government enacted law that provided ownership to an individual who cleared land and put it to effective use. Subsequently, *effective use* translated into clearing the

Figure 9.17 — The deforestation of Rondônia, Brazil, illustrates how anthropogenic drivers operate and effect landscape-use change. Rondônia is a state in Brazil, located in the northwestern part of the country. It is a rainforest area of the southern Amazon Basin that has undergone extensive deforestation beginning in the early 1970s. This 1985 satellite image illustrates the effects of anthropogenic change resulting from slash and burn agriculture (NASA image).

Landscape Change | 219

Figure 9.18 — Landscape-use change typically is preceded by various types of infrastructure development. The exploitation of Rondônia rainforest followed road corridor construction that provided the portal of entry for immigration (photograph by D. Gunzelmann).

rainforest landscape, principally for agricultural production and livestock grazing. The basic *commodities* (see the definition above) derived by the new landowner included pastureland and cattle, cropland and food, lumber for marketing, and firewood for cooking and comfort. Landownership *per se* was likely a social commodity. Development could not take place without provision of new infrastructure, and in Rondônia the principal precursor to landscape-use change was road corridor construction, which provided access for immigrants. Landscape-use change typically is preceded by various types of infrastructure development, and in addition to a road corridor network, other common precursors include electrification, health services, potable water, etc. Clearing the rainforest landscape is placed-based activity, and a general pattern emerged as a consequence of the road corridor exploitation (Figure 9.18). The basic method of landscape clearing was slash and burn, and this approach resulted in fields for agriculture and pasture for livestock. The unfortunate reality of agriculture on the thin soils of the tropical rainforest landscape is that it is not sustainable. Soil fertility is rapidly depleted and the spiral of succession is predictable with certainty: landscape clearing →agriculture →grazing →land abandonment →landscape clearing. In this example the anthropocentric driver is human response to economic opportunity made possible by institutional policy (Lambin et al. 2001, Veldkamp and Lambin 2001).

EFFECTS OF LANDSCAPE CHANGE ON LIVING ORGANISMS

In this section we return to an emphasis on ecology in landscapes. In particular, we consider how landscape-cover change and landscape-use change influence the habitat of living organisms. Recall that the term *habitat* was defined as the physical place where an organism either actually or potentially lives (Chapter 7). In the context of the transport model (Figure 8.2), the salient issue in this discussion centers on the consequences of habitat change to the persistence, distribution, and abundance of living organisms in the landscape environment. We examine three aspects of effects of landscape change on living organisms. First, we consider how landscape-use and landscape-cover change translates to the composition and geometry of the landscape. Second, we consider how habitat modification in landscapes affects living organisms. Third, we consider how landscape composition and geometry affect the demography of living organisms.

Change Effects on Landscape Composition and Geometry

Our focus here is directed to the spatially explicit result of landscape-cover and landscape-use change, i.e., how the various mechanisms of change directly translate to the composition and geometry of landscapes. Much of the ecological interest in and literature on this subject is bundled in discussion of landscape fragmentation and effects on habitat modification (Haila 2002, Fischer and Lindenmayer 2007). The general term *fragmentation*, meaning to break into smaller pieces, does not adequately cover the range of ways that change alters landscape structure. Forman (1995) provides a more robust vocabulary that includes the following: perforation, dissection, fragmentation, shrinkage, and attrition (Figure 9.19).

Perforation: The making holes in a landscape (Figure 7.8a).

Dissection: Separating or cutting a landscape into pieces (Figures 7.31 and 7.40).

Fragmentation: Breaking up a landscape into smaller pieces (Figure 7.11). Both fragmentation and dissection result in the creation of more

Spatial processes		Patch number	Average patch size	Total interior habitat	Connectivity across area	Total boundary length	Habitat Loss	Habitat Isolation
	 Perforation	0	-	-	0	+	+	+
	 Dissection	+	-	-	-	+	+	+
	 Fragmentation	+	-	-	-	+	+	+
	 Shrinkage	0	-	-	0	-	+	+
	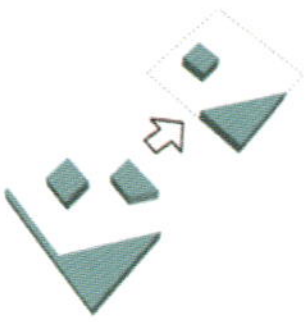 Attrition	-	+	-	0	-	+	+

Figure 9.19 — The mechanisms of landscape-cover and landscape-use change directly translate to the composition and geometry of landscapes through spatially explicit perforation, dissection, fragmentation, shrinkage, and attrition (modified from Forman 1995).

and smaller landscape elements. The distinction is that in a fragmented landscape, the pieces are more broadly dispersed and distributed. For fragmentation, imagine the dispersion pattern of shards that would follow from dropping a clay vessel onto a hard surface and shattering it. For dissection, imagine a landscape divided by the various types of corridors, e.g., roads, windbreaks, powerlines, etc. (Forman 1995).

Shrinkage: Decreasing the size of selected elements within the landscape (Figure 7.29).

Attrition: The loss of elements in the landscape. This type of landscape change is typically associated with conversion of landscapes from one use to another, e.g., replacing a prairie landscape with agricultural patches.

The different ways that landscape transformation takes place were defined above in the context of a spatially explicit geographical area consisting of characteristic component entities (ecosystems), i.e., a landscape. The terms can be used to describe change in the individual ecosystem components of a landscape, e.g., habitat patches (Figure 9.20). Furthermore, the terms can be applied to a cluster of interacting ecosystems that together form the functional habitat of living organisms. This last approach is examined in more detail below. In the case history example involving honey bees (*A. mellifera*) in pine forests (Chapter 1), the various landscape transformation processes were used to explain why this insect occurred in what should have been an inhospitable landscape.

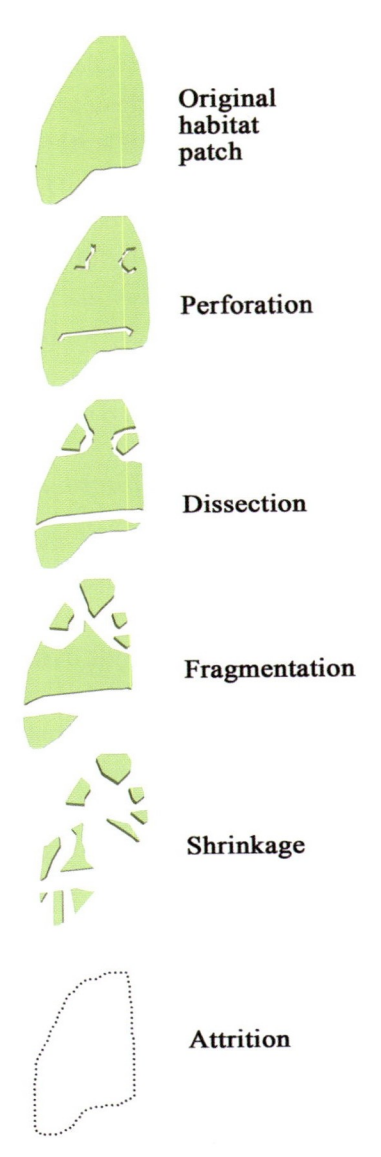

Figure 9.20 — The terms perforation, dissection, fragmentation, shrinkage, and attrition can also be applied to the individual ecosystem components of a landscape, e.g., habitat patches. This use is common in studies dealing with how fragmentation of habitat alters species persistence, distribution, and abundance as well as diversity (KEL drawing).

Landscape Change | 223

A suite of landscape pattern indices (LPIs) and spatial statistical procedures have been developed to characterize and compare change in landscapes (McGarigal and Marks 1995, Fortin and Dale 2005). The analytical procedures deal specifically with quantifying change resulting from perforation, dissection, fragmentation, shrinkage, and attrition of the basic elements of landscape geometry and composition (see Figure 2.10 and Table 2.1). The utility of the analytical procedures, in a landscape ecological context, is that they facilitate evaluation and comparison of the effects of landscape modification on habitat of living organisms. In Chapter 10 we examine metrics used for quantitative study of landscape structure, function, and change.

Habitat Modification and the Effects on Living Organisms

With our background on the mechanisms of landscape change, we can now examine the effects of habitat modification on species living in the landscape environment. The landscape transformation processes directly affect habitat of living organisms by reduction, degradation, and sub-division/isolation. Loss of critical habitat is primarily a function of landscape domestication practices. Along with the effects of invasive species, habitat loss is the principal cause for reduction in species diversity in landscapes. Degraded habitat results from a change in the conditions and resources of the landscape environment and is manifested in living organisms as reduced survival, growth, and/or reproduction (Figure 7.2). Sub-division/isolation is the breaking apart of suitable habitat into multiple units that are dispersed within the landscape. The extent to which landscape modification results in habitat sub-division/isolation is in large part a function of the movement capabilities of the species. Maintenance movement, dispersal pathways, and (in some cases) migration routes can be altered by habitat sub-division and isolation.

Clearly, habitat reduction negatively affects persistence, distribution, and abundance of living organisms in the landscape environment. Each species utilizes the landscape environment in unique ways. Therefore, no basic rules can define how much habitat is needed or how fragmentation will affect a given species a *priori*. This type of information is the substance of research in landscape ecology. Literature on the subject of habitat space (spatial extent) and the effects of fragmentation on extinction rates has been summarized by Fahrig (2002). Based on simulation modeling experiments (Fahrig 2002, Flather and Bevers 2002), a general relation describing habitat size and probability of persistence for a species was defined (Figure 9.21). Essentially, there is a minimum amount of habitat needed by a species, and below this threshold the probability of long-term population survival is less than 1.

Species interactions within the landscape environment are also altered by habitat modification. The direct effects take place through food web

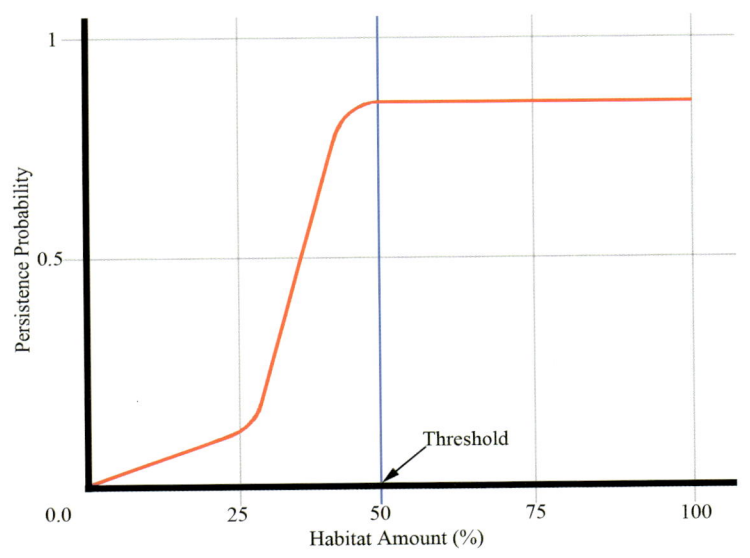

Figure 9.21 — The general relation of habitat size and probability of persistence for a species. Essentially, there is a minimum amount of habitat needed by a species, and below this threshold the probability of long-term population survival is less than 1. This relation is hypothesized from modeling experiments (modified from Fahrig 2002).

and trophic relations and are expressed as changes in processes such as predation, parasitism, competition, herbivory, and mutualism. To some degree, each species in a community is affected differently by landscape modifications associated with habitat loss, degradation, and sub-division/isolation. Niche requirement define the effects of landscape modification on individual species (Figure 7.3).

Landscape Change | 225

The Effects of Landscape Composition and Geometry on Living Organisms

How spatially explicit changes in landscape composition and geometry influence the persistence, distribution, and abundances of living organisms in the landscape environment is a central theme in landscape ecology research, natural resource management, and conservation biology. In this section, we examine landscape structure and habitat quality and quantity and consider the effects of change in this relation on the demography of living organisms. In particular, we consider four aspects of the interaction: habitat heterogeneity and species abundance, edge effects, landscape connectivity, and matrix effects. The literature on this subject has been reviewed by Ries et al. (2004), Tews et al. (2004), Harper et al. (2005), and Fischer and Lindenmayer (2007).

Habitat Heterogeneity and Species Abundance

In the context of landscape structure, we defined the term *heterogeneity* to be composition of parts of different kinds. In a similar way, *habitat heterogeneity* is defined to be the different places in a landscape suitable for occupancy and use by a living organism, i.e., the variety of physical places where an organism either actually or potentially lives. These habitat places can include patches, corridors, the matrix, and ecotones. However, most references in the landscape ecological literature are to habitat patches. This use of the term can be a generic reference to a landscape element or specific to our landscape ecological definition of a patch.

In terrestrial environments, landscape structure translates to animal habitat largely as a function of plant community organization. In general, there is a positive relationship between vegetation-based habitat heterogeneity and animal species diversity. Evidence supporting this relation was summarized by Tews et al. (2004) based on an examination of studies from different landscapes (forest, agricultural, prairie/steppe/grassland, scrub/shrubland, mountain, wetland, semi-desert, mangrove, and salt marsh) and a variety of taxa (arthropods, birds, mammals, amphibians, and reptiles).

Species diversity in vegetation-based habitat is also related to the patch size. The species-area relationship is a basic theme in landscape ecological research, and in general large patches support higher species diversity than smaller patches. Several interpretations are given to explain this relationship: Larger patches may have a higher ratio of colonization to extinction, they are more likely to contain undisturbed areas required by some species, they are more likely to include a broad range of environmental conditions that can accommodate a variety of taxa, and they are more likely to capture patchily distributed species by chance (Fischer and Lindenmayer 2007). Small patches can also serve as habitat and they further function as stepping stones that provide connectivity among the landscape elements. This topic is discussed below.

In summary, species diversity in the landscape is related to habitat heterogeneity and patch size. The various mechanisms of landscape change (perforation, dissection, fragmentation, shrinkage, and attrition) have greatly reduced native vegetation worldwide, both in variety (heterogeneity) of plant communities and their size (fragmentation). Reduced habitat heterogeneity and fragmentation diminish species diversity.

Edge Effects

Recall that an *ecotone* is a transition area occurring at the interface of two or more distinct landscape elements, e.g., patch types (Chapter 7). The adjacent components of an ecotone are referred to as edges (Figure 7.41). *Edge effects* (= edge influences) are changes in the resources and conditions of the ecotone that result from the adjacency of the landscape elements. The edges create an ecotone environment that is different from that of the adjacent landscape elements. Abiotic edge effects are changes in the physical state of the ecotone resulting from radiation, moisture, temperature, humidity, wind speed, soil nutrients, etc. Biotic edge effects are changes in the ecotone associated with living organisms, and include variables such as species abundance, diversity, and interaction (competition, predation, parasitism, herbivory, mutualism) (Ries et al. 2004, Fischer and Lindenmayer 2007).

The magnitude and distance of edge effects are a function of contrast in structure and composition between the adjacent landscape elements, e.g., the contrast between a forest and prairie grassland would be greater than that occurring between a natural prairie grassland and a planted coastal bermuda pastureland (Figure 9.22). In general, high contrast edges have a greater influence on landscape function than those with low contrast. In the landscape ecological literature, high contrast edges are often referred to as *hard boundaries* while low contrast edges are referred to as *soft boundaries*. Furthermore, the ecological significance of edge effects is related to the degree to which the ecotone habitat differs from the interior habitat of the adjacent landscape elements (Harper et al. 2005). Species adapted to edge habitats are referred to as *edge species* in contrast to those that utilize interior habitats, which are referred to as *interior species*. Some species utilize both edge and interior habitats.

In summary edges influence living organisms in several ways. Edges can amplify, attenuate, or deflect movement of individual organisms. Edges provide access to resources and condition of the adjacent ecological systems. Edges delineate resources and conditions that influence and thereby define the distribution of different organisms. Edges provide unique habitat relative to the adjacent landscape elements. Edges serve

Hard Boundary

Soft Boundary

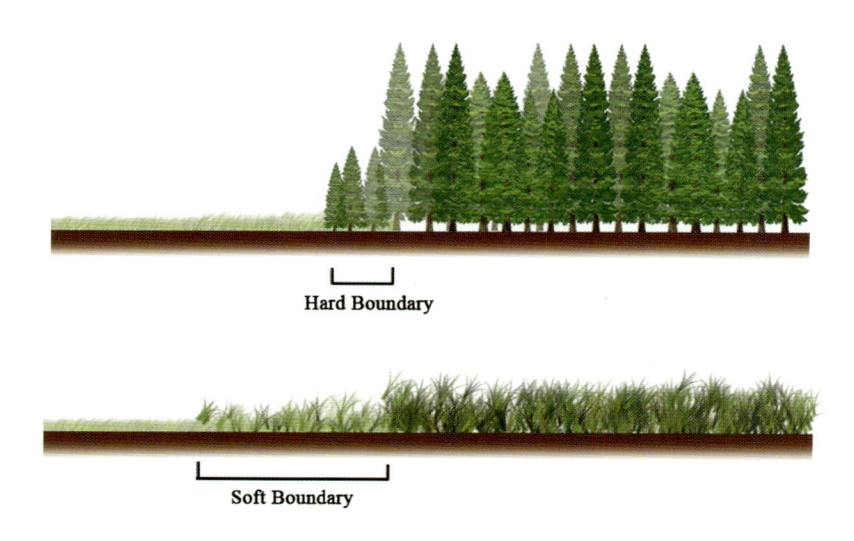

Figure 9.22 — Edge effects (= edge influences) are changes in the resources and conditions of an ecotone that result from the adjacency of the landscape elements. The edges create an ecotone environment that is different from that of the adjacent landscape elements. The contrast between adjacent edges results in hard boundaries (a) and soft boundaries (b) (KEL drawing).

as filters for movement of energy, materials, and information. All changes in habitat heterogeneity and patch size (discussed above) are passed to the edge environment of the adjacent landscape elements, and therefore can directly affect the species abundance and diversity in either positive or negative ways. Within a specific landscape, new edges can result from unique patch associations caused by the change mechanisms, and natural edges can be eliminated.

Landscape Connectivity

The concept of connectivity is founded on knowledge of how landscapes are structured (Chapter 7) and how the organization of the basic elements affects movement of energy, materials, and information (Chapter 8). Clearly, the geometry and composition of the landscape (i.e., the patch, corridor, matrix, ecotone configuration) influences the movement of all entities within the landscape environment. In this section our interest is on how change in landscape structure affects individuals as well as communities of living organisms.

Recall that *functional heterogeneity* was defined to be a perception-based concept, i.e., how an organism perceives and responds to the landscape environment. In assessing the suitability of a landscape for individual species and in evaluating the impact of change, we are interested in knowing if the requisite habitat patches are present, if they are of sufficient size, and if they are accessible. The first two points were examined above. An additional component of functional heterogeneity, therefore, is *habitat connectivity*, the connectedness between patches of suitable habitat in the landscape (Fischer and Lindenmayer 2007). This attribute of the landscape environment is defined by maintenance movement behavior and dispersal capabilities of the species in relation to the spatial locations of the habitat patches. Together the presence, sufficient size, and accessibility of suitable habitat patches establish *functional connectivity* of the landscape for an individual species. Both of these concepts (habitat connectivity, and functional connectivity) apply to migration routes that involve a much larger spatial extent than encompassed in a landscape or ecosystem cluster. The functional connectivity of a landscape can be changed, in reference to an individual species, by reducing the presence of a requisite habitat patch, reducing the size of the patch below a threshold of utility, or isolating the patch. Again, landscape connectivity influences community interactions (predation, parasitism, competition, herbivory, and mutualism) through food web and trophic interactions.

Connectivity is a landscape attribute that is often manipulated for management purposes. Examples of such actions include the construction of windbreaks to diminish soil erosion, dams to control flooding, wildlife corridors to facilitate protected movement of species in fragmented landscapes, etc. (Figure 9.23) These landscape-use changes can disrupt or enhance landscape connectivity.

The Matrix

Recall that the *matrix* is the background or cover type of a landscape and that this element is generally distinguished from the other components by area (spatial extent), connectivity, and control over dynamics. Historically, landscape ecologists have viewed the matrix simply as a place holder for patches, corridors, and ecotones. However, there are numerous, matrix-dominated landscapes, e.g., deserts, boreal forest, rain forests, tundra, etc. Furthermore, the matrix is habitat to a variety species of living organisms, e.g., the ungulates populating the grasslands of Kenya and Tanzania, Africa. The effect of the landscape transformation processes on matrix habitat is to add heterogeneity into the homogenous environment. In this circumstance increasing heterogeneity may not be a desirable landscape change and it can have profound negative effects on matrix-adapted species (Figure 9.24). The landscape transformation processes directed to the matrix are often the result of landscape-use activities, e.g., conversion of grassland to agricultural patches, harvesting forests, extracting minerals, etc. Each activity results in fragmentation of the homogeneous matrix environment.

DEVELOPMENT OF PATTERN IN MOSAIC LANDSCAPES

In the preceding sections of this chapter, we have examined how landscape-cover change and landscape-use change result in modification of landscape structure. We further considered how changes in landscape structure translate directly to habitat of living organisms and the consequential effects on persistence, distribution, and abundances of resident species. However, the question remains: How does landscape-cover change and landscape-

Figure 9.23 — Connectivity is a landscape attribute that is often manipulated for management purposes. Examples of such actions include the construction of windbreaks to diminish soil erosion, dams to control flooding, wildlife corridors to facilitate protected movement of species in fragmented landscapes, etc. This wildlife corridor facilitates safe movement of species across a road corridor (KEL drawing).

Landscape Change | 231

use change create observed patterns in landscapes? Following, we address this question, and have partitioned the discussion into two compartments: pattern in natural landscapes and pattern in human dominated landscapes.

Pattern in Natural Landscapes

In some measure, all landscapes have been and are being shaped by human activities. The effects can be the direct result of atmospheric deposition of the byproducts from industrialization (e.g., acid rain, photochemical oxidants, etc.) or due to changes in temperature and precipitation resulting

Figure 9.24 — The effect of the landscape transformation processes on matrix habitat is to add heterogeneity into the homogenous environment. In this circumstance increasing heterogeneity may not be a desirable landscape change and it can have profound negative effects on matrix-adapted species (NASA image).

from anthropogenic-induced global climate change. Although the initiating causes are widely distributed in origin, the effects on the landscape are site- and species-specific.

In heterogeneous natural landscapes, the observed pattern is a function of the underlying landform, climatic and edaphic conditions, disturbance regime, activities of living organisms, and cumulative historical events that have taken place over time. How these processes play out in specific landscapes, i.e., result in spatial patterning, is one of the major research arenas of landscape ecology. There is no general set of assembly rules that can be used to describe how landscape pattern occurs. However, detailed examination of the processes, enumerated above, can be directed to specific landscapes. Examples of the application of this approach are provided by van der Valk and Wagner (2009) for landscapes where the pattern is influenced by wetlands and for arid mosaic landscapes.

Of equal interest to the landscape ecologist is the subject of how the landscape mosaic changes through time. The ecology literature is laden with the reference to the "shifting mosaic" of ecosystems that form a landscape. The implication is that there is an orderly progression, perhaps an adaptive cycle as illustrated in Figure 4.8, for change in landscape pattern. This sequence would be reflective of the processes that cause change. The time frame (temporal extent) for ecological succession of landscape mosaics exceeds historical records. Consequently, we are left with observing a few scenes extracted from an unknown reference point in the cycle of change.

Pattern in Human-Dominated Landscapes

Landscape-use change is the principal cause of landscape-cover change. The basic human needs are the motivators, but response to economic opportunity, mediated through governmental institutions, often drives landscape-cover change (Lambin et al. 2001). The principal human activities that result in landscape-cover change are associated with agriculture, forestry, grazing, urbanization, and commerce. Each produces a pattern within the landscape that is unique to the activity. The specific causes for the patterns produced are the subject domain of forestry, agronomy, range management, and urban design. However, there are generalizations that apply to each of the systems, including the following: angular geometry, energy subsidies, and historical landscape-use. Each topic is discussed below.

Angular geometric pattern is a characteristic of human-manipulated landscapes. Each of the activities that result in landscape-cover change (identified above) introduces angular geometry to the landscape. The pattern occurs for a variety of reasons: landownership, political boundaries, economy of scale (practical limits for managing an agricultural field or forest stand), efficiency in movement (corridor structure), aesthetic appearance (design in parks), etc. When you view an aerial photograph your eyes are drawn to geometric pattern, and this feature is strong indicator of human presence in the landscape (Figure 9.25).

All landscape-use changes involve anthropogenic energy input. Furthermore, maintenance of the structure in human-manipulated landscapes requires continuous energy input, as dictated by the second law of thermodynamics. Without energy input, entropy increases and order decreases and eventually the human-generated structure dissipates. For example, an agricultural field (an introduced patch) would soon be absorbed into the parent landscape if tillage, weed control, harvesting, and replanting practices were abandoned. Each of these activities involves an energy subsidy. Energy subsidies in urban landscape swamp all of the other landscape-use activities. Although the area occupied is small, relative to the other landscape uses (about 2 percent of the earth's land surface), the ecological footprint is large (Stöglehner 2003).

Human settlement historically clustered in hospitable landscapes, i.e., places where the human needs could be satisfied. As plant and animal domestication practices advanced, landscapes where climatic and edaphic conditions were favorable for agricultural production were exploited, in particular (Terrell et al. 2003). Generally, these places have been utilized for similar purposes for many generations. In effect, this overprinting is a palimpsest that retains features of both the original and current landscape (Thomas 2001). Given the worldwide distribution of domesticated food plants and animals, the specific flora and fauna may differ in detail from the original landscape. However, the biota will have similar adaptations to the landscape environment as the original organisms (Figure 9.26) (Haines-

Figure 9.25 — Angular geometric pattern is a characteristic of human-manipulated landscapes. Each of the activities that result in landscape-use change introduces angular geometry to the landscape. This figure illustrates a Balinese Subak (agricultural community) (photograph by Adriano/Wikimedia Commons).

Landscape Change | 235

Young 2005). In evaluating the observed pattern in landscape structure, historical human use is a component as important as the effects attributed to the other change processes (Antrop 2005, Werner and McNamara 2007).

EPILOGUE

So what conclusions can be made about landscape change?

1. *Landscape change* deals with the alteration of structure and function of the landscape environment over time. This subject is broad-based and includes a variety of topics that are of fundamental importance to landscape ecologists and practitioners of landscape management.

2. As a cautionary note, much of literature on the topics covered in this chapter originated at the early stages in the genesis of landscape ecology, and consequently, the terms *landscape* and *ecosystem* were often used interchangeably, imprecisely, or as a metaphor for holism. In this chapter we have been steadfast in using the scientific definitions of landscape and ecosystem, as defined in Chapters 1 and 4.

3. The subject of landscape change was broken into four parts: landscape-cover change, landscape-use change, how landscape change affects animal habitat, and how change is reflected in landscape pattern.

4. Landscape-cover change deals specifically with the alteration of biophysical attributes of the landscape environment. We examined three aspects of landscape-cover change: biogeomorphology, the activities of living organisms, and the concept of disturbance. The discussion of biogeomorphology emphasized the interplay of landforms and living organisms. The discussion of the role of living organisms in effecting landscape change centered on the activities of ecosystem engineers, keystone species, and invasive species. In addition to the roles that living organisms play in landscape-cover change, we also considered the concept of disturbance. A disturbance was defined to be an initiating cause that produced an effect greater than average, normal, or expected. We examined how disturbances impact landscapes and introduced the concept of disturbance regime.

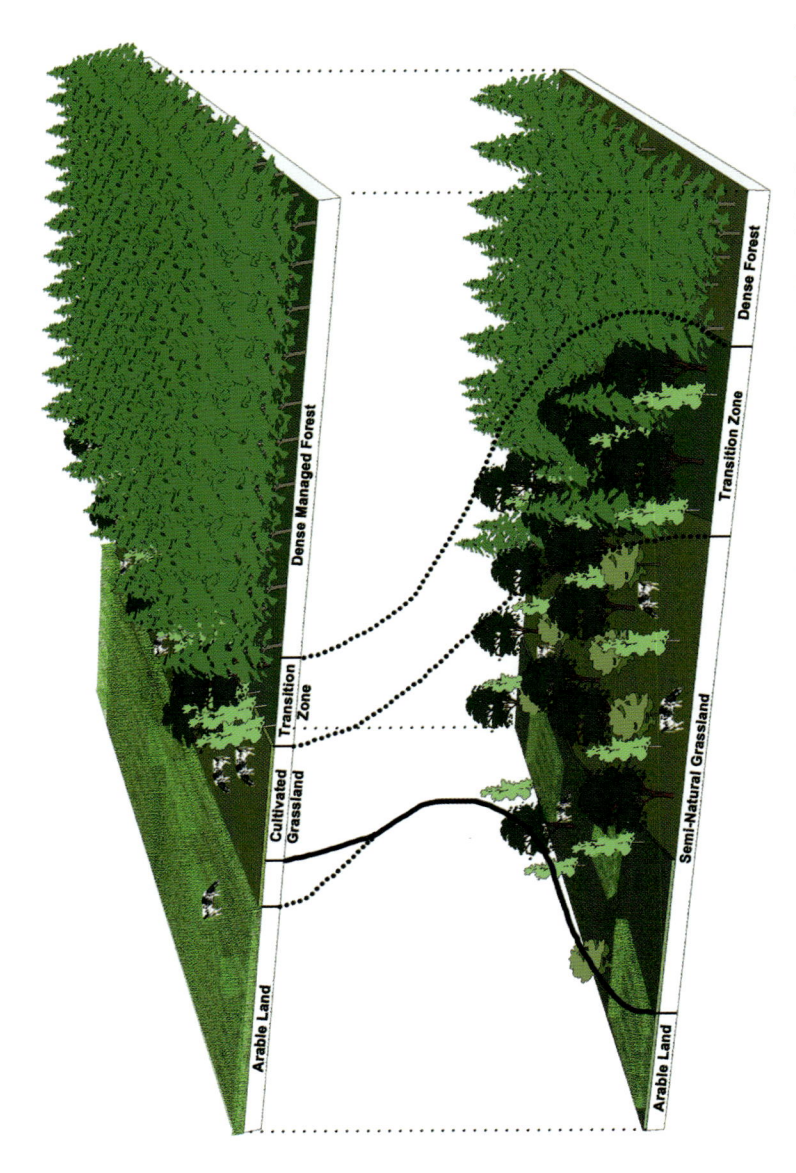

Figure 9.26 — Human settlement historically clustered in hospitable landscapes, i.e., places where the human needs could be satisfied. Generally, these places have been utilized for similar purposes for many generations. In effect, this overprinting is a palimpsest that retains features of both the original and current landscape (modified from Haines-Young 2005, and based on the original illustration by Skånes).

Landscape Change | 237

5. Landscape-use change deals with human purpose or intent as applied to the biophysical attributes of the landscape environment. This type of change has the greatest impact on landscapes and was brought about by human involvement in agriculture, natural resource management, range management, and urbanization. We organized this subject within the concept of landscape domestication, i.e., the activities of humans that structurally shape and functionally modify landscapes to satisfy basic human needs. The basic human needs were presented as the drivers of change in landscape-use. A driver was defined simply to be a human-initiated cause (physical force, a process, or an event) that results in landscape-use change.

6. The effects of landscape change on living organisms is a fundamental issue in landscape ecology and a subject of scientific inquiry, landscape management, and conservation biology. Emphasis was placed on how landscape-cover change and landscape-use change influence the habitat of living organisms. In the discussion, we examined how change actually affects landscape composition and geometry through processes of perforation, fragmentation, dissection, attrition, and shrinkage. How these processes result in habitat modification and effects on living organisms were examined. Landscape composition and geometry also have direct affects on living organisms and we considered four aspects of the topic: habitat heterogeneity and species abundance, edge effects, landscape connectivity, and matrix effects.

7. The final topic in this chapter addressed the issue of how landscape-cover change and landscape-use change create observed patterns in landscapes. We examined pattern formation in natural landscapes and human-dominated landscapes.

10

Landscape Analysis and Synthesis

OVERVIEW

The purpose of this chapter is to examine how data, information, and knowledge of landscape ecology can be used in scientific investigations and for landscape-use management. Recall from Chapter 8 the definitions given for data, information, and knowledge. *Data* are the measurements that define an ecological phenomenon, process, or relationship of interest, i.e., a sonogram of a bird "song." *Information* is data that have been given meaning by way of relational connection, i.e., attributing a sonogram to a specific bird species. *Knowledge* is contextually integrated information, i.e., an ornithologist recognizing how the behavior of a bird species is influenced by the ensemble of different "songs" it makes.

Landscape ecology is, in part, a computational science. The digital age facilitated and inspired the development of a myriad of computer-based methods for spatial data analysis, synthesis of information, and the integration of knowledge. An understanding of the various methods and their appropriate uses is fundamentally important in interpreting the literature of landscape ecology and in using your knowledge of the subject for scientific investigations and in practical applications. Accordingly, our goal in this chapter is to provide a systematic overview of the various approaches used by landscape ecologists for spatial data analysis, synthesis of information, and integration of knowledge. We have three objectives. The first objective deals with an examination of the various procedures used in landscape data analysis. The second objective deals with synthesis of data and information using spatial modeling procedures. The third objective deals with procedures for integrating data, information, and heuristic knowledge of experts (Figure 10.1).

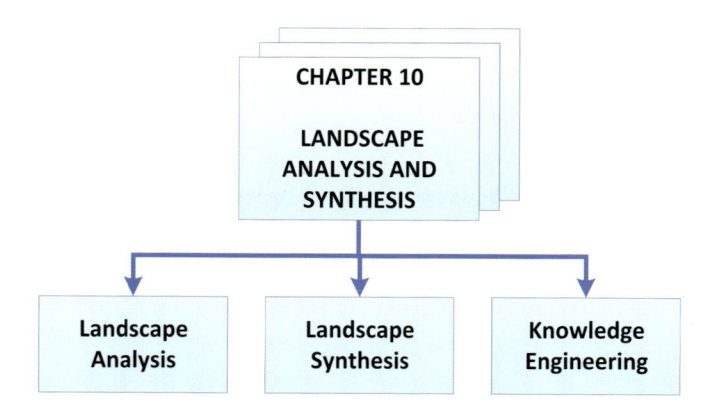

Figure 10.1 — Summary of topics and organization of Chapter 10.

ANALYSIS OF LANDSCAPES

Analysis deals with separating or breaking up of any whole into its parts so as to find out their nature, proportion, function, or relationship. In a landscape, the "whole" is represented as heterogeneity. In landscape ecology, we are interested in analysis of spatial data describing the constituent elements that form the landscape mosaic and in the measurements that define ecological phenomena, processes, and relationships taking place within the landscape environment. In the case of the landscape mosaic, the P/C/M/(E) model provides for the formal representation of the compositional elements of heterogeneity. The data are often associated with categorical maps, and the measurements are summarized using landscape pattern indices (LPIs). In the case of landscape ecological phenomena, processes, and relationships; the measurements are point-data (represented by units and dimensions), and analysis involves application of spatial statistics. In the following sections, we revisit the concept of spatial heterogeneity (introduced in Chapter 7) and examine procedures for analysis of landscape pattern and point-data.

A familiarity with the basic analytical methods provides the means for integrating and interpreting your knowledge of landscape ecology concepts. Typically, a landscape ecologist will assemble a "tool box" that contains the variety of procedures needed to address subjects of personal interest, e.g., LPIs for characterization and comparison of landscape structure. The more richly supplied the tool box, the greater the range of approaches available for creative investigations of landscapes. By contrast, the opposite situation is true as well, and Maslow (2002) succinctly described the consequence as follows: "I suppose it is tempting, if the only tool you have is a hammer, to treat everything as if it were a nail." Furthermore, knowing how to use a tool is as important as choosing the proper one.

Heterogeneity

The concept of *heterogeneity* is a pervasive theme in landscape ecology. Throughout this text, we have used the term *heterogeneity* to mean composition of parts of different kinds. Whereas this definition is suitable for many purposes, further elaboration is needed when the term is used in the context of spatial analysis, i.e., when we disassemble and quantitatively examine the basic elements of a landscape or investigate an ecological phenomenon, process, or relationship. The following definition is tailored to capture this meaning: *heterogeneity* – "the spatially structured variability of a property of interest, which may be a categorical or quantitative, explanatory or dependent variable" (Wagner and Fortin 2005). The quantification of spatial heterogeneity requires a way to describe and represent variability in space and time (Gustafson 1998), and this activity is accomplished through the application of the LPIs and spatial statistics.

In Chapter 1 we defined landscape ecology to be the science that embraces the agenda of ecology in a spatially explicit manner. This point of emphasis is amplified when the approaches used for analysis in ecology are contrasted to those used in landscape ecology. Figure 10.2 is a schematic representation of the conceptual framework for ecological (A) and landscape ecological analysis (B). In the ecological framework, the ecological processes (the upper surface in A) observed in a set of plots (white squares) depends on the nature of the environmental factor (polygons in the lower surface) measured at the plot locations. The patches/plots are internally homogeneous, plot context does not matter, and observations are spatially independent. In the landscape ecological framework, the patches/plots may be internally heterogeneous, plot context may affect local processes, and observations may not be independent due to spatial interactions between local processes (Wagner and Fortin 2005).

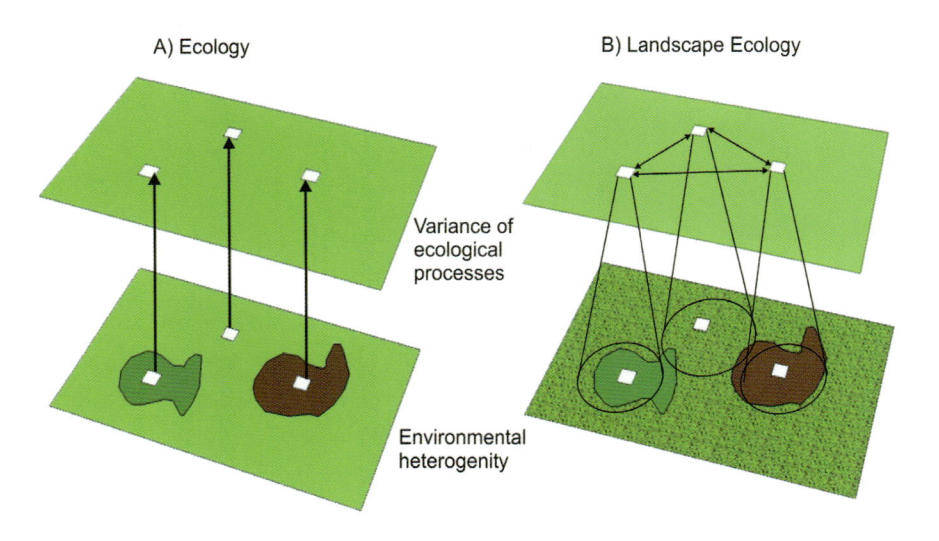

Figure 10.2 — A conceptual framework for ecological and landscape ecological analysis. In A (the ecology framework) the patches/plots are internally homogeneous, plot context does not matter, and observations are spatially independent. In B (the landscape ecological framework) the patches/plots may be internally heterogeneous, plot context may affect local processes, and observations may not be independent due to spatial interactions between local processes (modified from Wagner and Fortin 2005).

The ecological consequences of landscape heterogeneity in relation to individual species are manifold. Basic generalities include the following. Habitat patches linked by dispersal tend to have interdependent population dynamics, which leads to autocorrelation in species abundance. Landscape-use practices can create a mosaic that separates habitat patches, and thereby introduces spatial structure in species abundance. Functional connectivity of habitat patches for a species can be disrupted by landscape-use practices. Functional heterogeneity of the landscape mosaic is species-specific and dependent on behavioral adaptations associated with life history attributes (Wagner and Fortin 2005).

Analysis of Landscape Pattern

A large number of LPIs have been developed to quantify spatial heterogeneity associated with categorical maps (Figure 10.3). The various approaches were initially assembled by McGarigal and Marks (1995) in FRAGSTATS. This software program is designed to compute a variety of landscape metrics for categorical map patterns. Contemporary applications and instructions for using the system are documented at <http://www.umass.edu/landeco/research/fragstats/fragstats.html>. The utility, suitability, and limitations of the LPIs for landscape analysis have been reviewed by Gustafson (1998), Liebhold and Gurevitch (2002), Li and Wu (2004), and Wagner and Fortin (2005).

The categorical elements that create the heterogeneity in a landscape mosaic have four attributes that are important in landscape analysis: composition, configuration, shape, and scale. *Composition* deals with the number and relative frequency of a landscape element (ecosystem). *Configuration* deals with the spatial arrangement (juxtaposition) of the basic landscape elements (ecosystems) in relation to one another. Emphasis can be directed to individual ecosystems or to clusters (neighborhoods). *Shape* deals with the geometry of the individual elements (ecosystems) and brings into play edge effects. Scale has spatial and temporal components. *Spatial scale* deals with the grain (resolution) and extent (range) of the landscape mosaic (Chapter 3). *Temporal scale* deals with the states of change in a landscape mosaic through time. For practical purposes the landscape elements (ecosystems) are generally referred to in the literature as patches. But where the patches are located in the matrix and how they are connected by corridors are important considerations in landscape analysis. Furthermore, categorical analysis requires user-based decisions regarding the degree of variation that will be permitted within a patch, the minimal size of patches that will be included (minimal mapping unit), and the components of the system that are ecologically relevant to the organism or process of interest (Gustafson 1998). Figures 4.2 and 5.2a and b illustrate composition, configuration, and shape. Figure 7.40 illustrates the issue of spatial scale. The case history examples describing the interaction of the southern pine beetle (*D. frontalis*) and the Red-cockaded woodpecker (*P. borealis*) and habitat requirements for honey bees (*A. mellifera*) (Chapter 2) illustrate

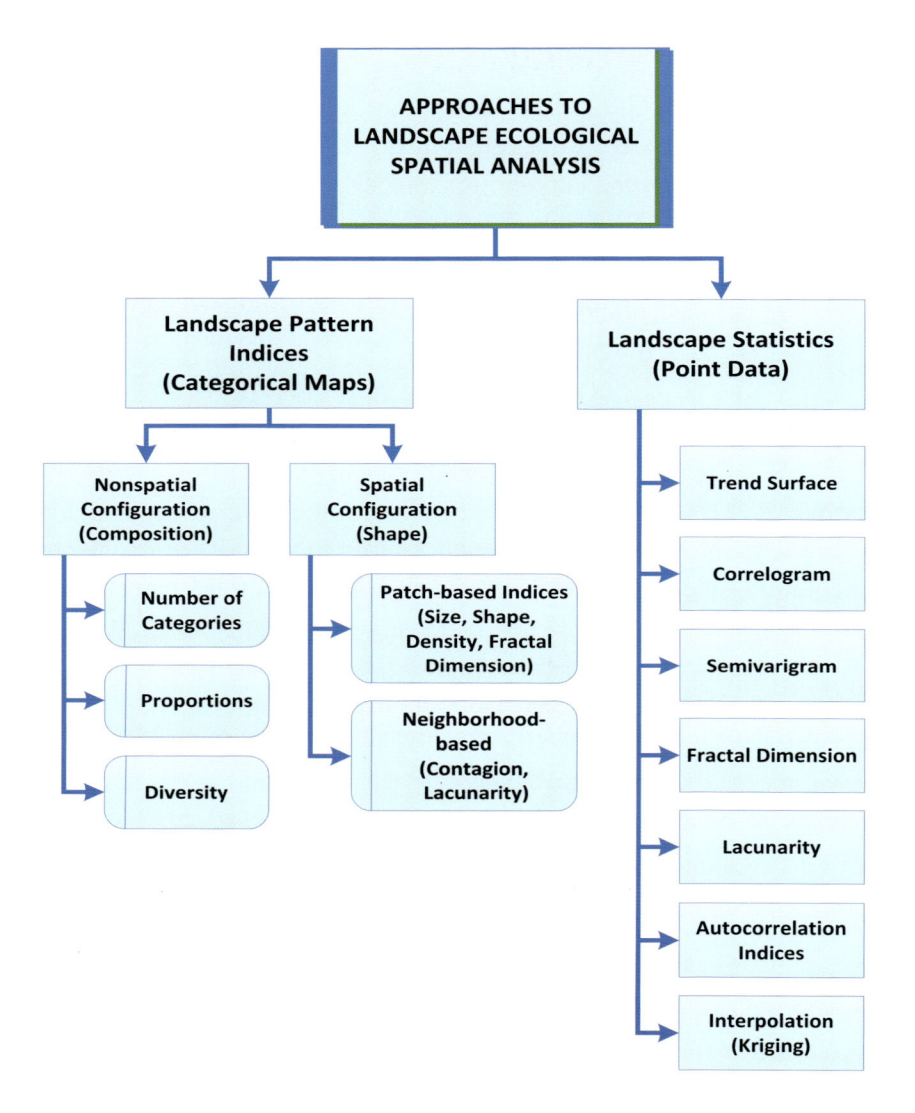

Figure 10.3— Systematic classification of methodologies used in landscape pattern analysis and landscape statistics (modified from Gustafson 1998).

the utility of LPIs in landscape ecological research. Much of the need for spatial pattern indices is driven by interest in predicting the response of some ecological entity (e.g., fire, organisms, or nutrient flux) to the spatial heterogeneity of managed landscapes (Gustafson 1998).

Analysis of Point-Data in Landscapes

As with the data associated with the compositional components of the landscape, the quantitative measurements that characterize ecological phenomena, processes, and relationships are also spatially structured and vary through time. The measurements are represented as point locations in the landscape mosaic and can be associated the elements of the landscapes (e.g., habitat patches) or independent of a categorical classification. Point-data are collected by sampling. The actual data are characterized by units and dimensions, e.g., plant biomass could be represented as g/m^2. Typically, a landscape is sampled to generate spatially referenced data about a variable of interest, e.g., the distribution and abundance of the red imported fire ant (*S. invicta*) (see case history below). Analysis of point-data is accomplished using spatial statistics and flavors of the different techniques are listed in Figure 10.3. In the literature these procedures are often referred to as *geostatistics*. Spatial statistics (=geostatistics) provide a methodology to describe spatial structure and make inferences about spatially dependent relationships in landscapes. Point-data analysis assumes that the system property of interest is distributed in a spatially continuous manner (Gustafson 1998). Spatial statistics are extremely important in analyzing ecological phenomena, processes, and relationships in landscapes as they address the issue of spatial autocorrelation and independence of observations. Parametric statistical analyses are based on the assumption of independence among samples, and this circumstance is violated in most landscape ecological studies. Indeed, spatial dependency associated with ecological entities in patches and gradients in the landscape environment are often the issues of interest in landscape ecological investigations (Fortin and Dale 2005).

The following case history example illustrates the utility of spatial statistics in landscape ecological investigations. In this study, population dynamics of the red imported fire ant (*S. invicta*) was investigated. The goal was to develop a risk rating system, and to accomplish this end it was necessary to sample the distribution and abundance of the insect in a heterogeneous landscape. The study was conducted on a 275 ha post oak savanna landscape in Central Texas. Following the assembly of a thematic spatial database, the landscape mosaic was classified using a GIS and the Patch/Corridor/Matrix/(Ecotone) model (Figures 4.2, and 5.2a and b). A 200 x 200 m primary grid was created and superimposed on the landscape mosaic image. Each cell was assigned a unique number. Sample points were selected by subdividing the primary grid into a 10 x 10 m secondary grid. Coordinates for each sample location were selected randomly from this grid and used as GPS waypoints for navigation to the specific sites in the landscape (Figure 10.4). Ant mounds were used as a

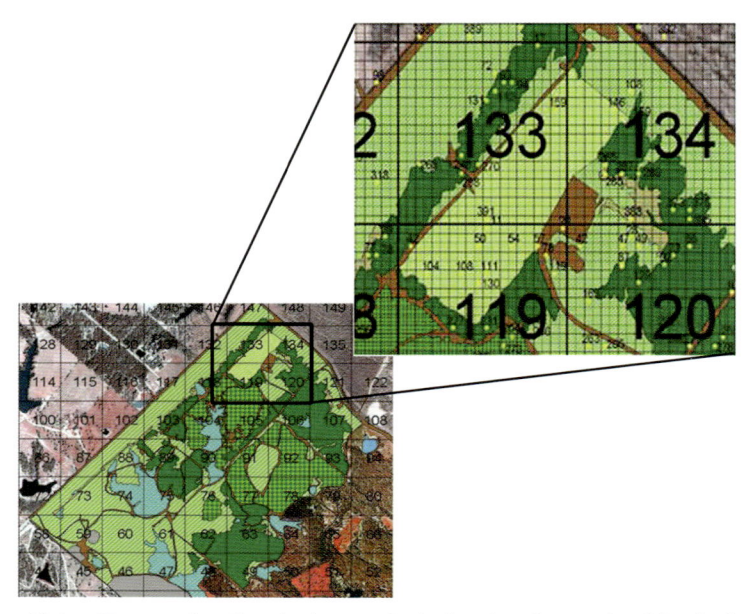

Figure 10.4 — The procedure for selecting sample site locations for a study of the distribution and abundance of the red imported fire ant (*S.invicta*) in a post oak savanna landscape. A 200 x 200 m primary grid was created and superimposed on a classified landscape mosaic image. Each cell was assigned a unique number. Sample points were selected by subdividing the primary grid into a 10 x 10 m secondary grid. Coordinates for each sample location were selected randomly from this grid and used as GPS waypoints for navigation to the specific sites in the landscape. Ant mounds were used as a surrogate variable to characterize population density. The plot size was 0.01ha, which translated to a 5.65m radius measured by a rope (see Figure 7.6) (KEL image).

surrogate variable to characterize population density. The plot size was 0.01 ha, which translated to a 5.65 m radius measured by a rope (Figure 7.6). The point-data (number of ant mounds/0.01 ha) were recorded using an electronic data logger, downloaded directly to a database in the GIS, and subsequently analyzed. The landscape was sampled in the spring and late summer each of two years. A result of the sampling exercise is illustrated in Figure 10.5. The statistical procedure used to summarize the point-data was a kriging technique which interpolated the measurements

Figure 10.5 — The surface trend in distribution and abundance of the red imported fire (*S.invicta*) occurring in a post oak savanna landscape. The statistical procedure used to summarize the point-data was a kriging technique which interpolated the measurements of ant density across the landscape mosaic (KEL image).

of ant density across the landscape mosaic. As the landscape mosaic was also classified, the data could be summarized categorically as well. Figure 10.6 illustrates fire ant density in different vegetation types associated with the post oak savanna landscape. The important message conveyed in Figure 10.6 is that the red imported fire ant has preferences in vegetation type, and therefore, the patch structure of the landscape influences where the insect will be found, i.e., there is a spatial dependency of the ecological entity associated with patches and gradients in the landscape.

Mounds per Hectare by Land-Cover Classification

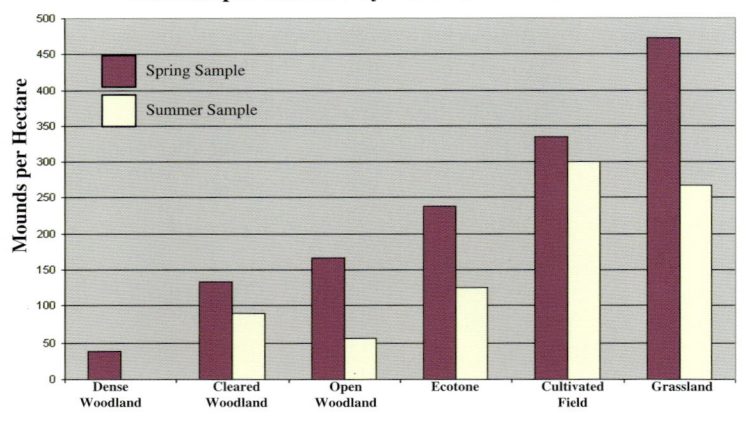

Figure 10.6 — As the landscape mosaic in Figure 10.5 was also classified by composition, the data could be summarized categorically. The bars in the histogram define fire ant density in different vegetation types associated with the post oak savanna landscape. The important message conveyed in this figure is that the red imported fire ant (*S. invicta*) has preferences in vegetation type, and therefore, the patch structure of the landscape influences where the insect will be found, i.e., there is a spatial dependency of the ecological entity associated with patches and gradients in the landscape (KEL image).

SYNTHESIS OF SPATIAL DATA AND INFORMATION IN LANDSCAPES

Synthesis deals with the putting together of parts or elements so as to form a whole, i.e., it is the antithesis of analysis. In landscape ecology a great deal of emphasis has been placed on synthesis involving the use of spatially explicit simulation models. In the following sections we provide a general overview of landscape modeling and examine a case history that illustrates the utility of the approach.

Spatially Explicit Landscape Simulation Models

Spatially explicit models are computer programs that incorporate the complexity of the real-world landscape in that they include both topological and chorological components. Simulation models that deal with ecological issues in landscapes are, *de facto, spatially explicit*, i.e., the entities associated with the model are defined by map coordinates and

248 | *Landscape Analysis and Synthesis*

placed within a geographical context. Geographic Information Systems (GISs) have greatly aided in the development and use of spatially explicit models as this software technology is typically used to input, store, and display data used in crafting the models and in their application (Figure 10.7). A great deal of the focus in landscape ecology has centered on *process* modeling. Recall that a process was defined simply to mean a change that leads towards a predictable result, e.g., herbivore consumption, elemental movement within a landscape, provision of ecosystem services, etc. Process modeling is "the representation of the dynamic behavior of a continuous 'real' physical or biological system over space and time" (Reiners and Driese 2004).

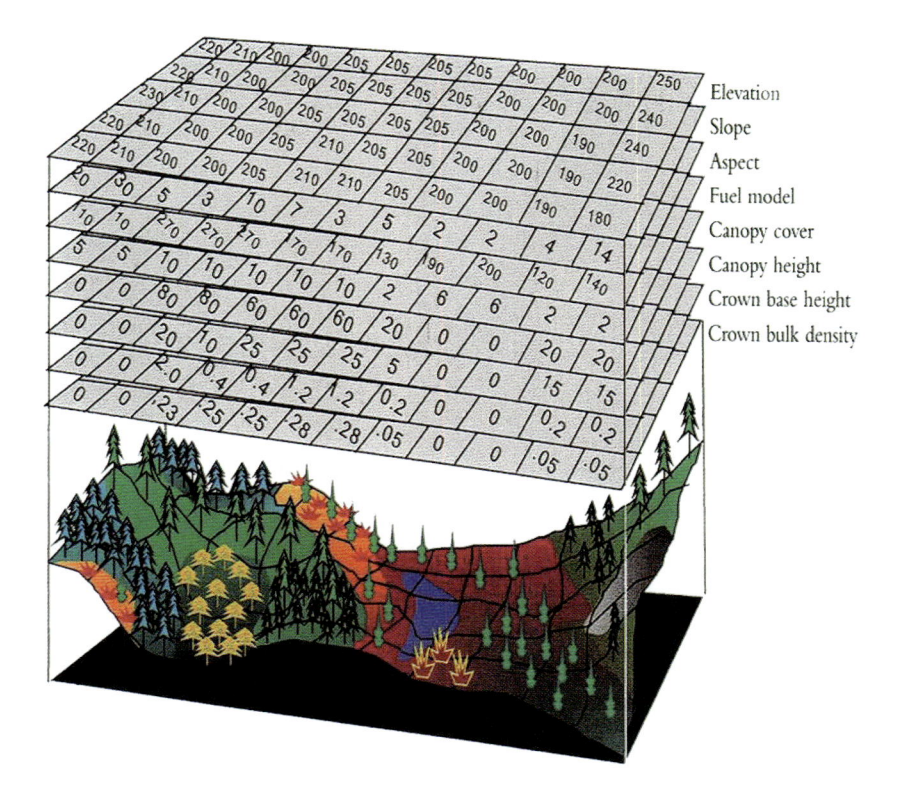

Figure 10.7 — Simulation models that deal with ecological issues in landscapes are spatially explicit, i.e., the entities associated with the model are defined by map coordinates and placed within a geographical context. Geographic Information Systems (GISs) have greatly aided in the development and use of spatially explicit models. This software technology is typically employed to input, store, and display data used in crafting the models and in their application (from Mladenoff and Baker 1999, figure by Mark Finney).

The approaches used in the development of spatially explicit models have been reviewed by Wu and David (2002), Perry and Enright (2006), Scheller and Mladenoff (2007), He (2008), Perry and Millington (2008), Gaucherel and Houet (2009), and Xi et al. (2009a). The spatial modeling enterprise has expanded with the development of computer languages that facilitate integration of diverse numerical and computational methods. In particular, object oriented design (OOD) is a methodology that has permeated the field. This approach facilitates the development of multi-purpose and multi-scale applications built from modular components (Scheller and Mladenoff 2007). All spatial process models require specification of four basic components: time, space, state variables, and interactions (Table 10.1). Spatial simulation models are used for a variety of purposes, e.g., prediction (effects of herbivory on tree mortality in a forest), projection (estimating population growth of an endangered species), exploration (examination of different planting regimes [scenarios] for optimization of forest growth), and explanation (illustrating how barriers in a landscape change species movement patterns) (Gustafson 2002). One of the principal applications of spatial modeling is in capturing the inherent complexity and dynamic behavior of landscapes. The complexity stems from the landscape heterogeneity, and the dynamics is a function of interactions that take place within and among the component ecosystems, i.e., the flux of energy, materials, and information. The interactions also occur across a range of spatial and temporal scales. One significant challenge in spatial modeling centers on defining the appropriate range and resolution needed for a particular study. Spatial interactions among the ecosystem components of landscapes produce emergent behavior and spatial patterning at multiple scales. The interactions, therefore, contribute to the evolution of landscape pattern and changes in spatial heterogeneity (Scheller and Mladenoff 2007).

An Application of Spatial Modeling –Forest Landscape Structure and Bark Beetle Herbivory

In the following example of spatial simulation modeling, we examine how modification of landscape structure affects population dynamics of the southern pine beetle *(D. frontalis)* (Figure 2.11). Some aspects of the natural history of this pest insect species are discussed in the case history

Table 10.1 — Basic components of spatial process models (KEL image - figure by Andrew Birt and Yu Zeng).

	Relevant Questions to Address	Examples
Spatial Scale	What area does the model cover, and at what resolution?	Spatial extent = 100 to 1000 m^2 Spatial Resolution < 1m^2
Temporal Scale	Over what time period does the model run, and what is the time-step?	Between 1 and 365 days on a daily time-step.
Number of State Variables	How many different *entities* are represented by the model?	**Weather** - Wind and temperature. **Beetles** - Development, survival, reproduction, dispersal, and attack behavior. **Trees** - Diameter, location, and mortality.
Number of Process Interactions	How many interactions are there between the different *entities* of the model? Is there 'feedback' between the different processes in the model?	**Weather** drives **beetles** (development, survival, reproduction, dispersal, and attack behavior) which drives **tree** mortality. No feedback between tree mortality and weather.

Notes:

1) Overall model complexity is some function of all these elements.

2) Summarizing a model in this way is the first step in determining whether it is fit for a specific purpose. For example, is it too complicated, are all the interacting processes understood and well defined, does it satisfy a research objective?

3) Modelers may approach model development by considering each building block sequentially in an order determined by the research objective, the availability of data, and the tools available to them.

4) Example derived from TAMBEETLE (Coulson et al. 1989).

example dealing with the interaction of beetles and birds in Chapter 2. Because *D. fontalis* is a pest species of great economic importance in pine (*Pinus* spp.) forests of the southern United States, it has been the focus of considerable research attention. One product of the research is a comprehensive biophysical model of population dynamics of the insect (Feldman et al. 1981), TAMBEETLE. The original computer code for this model was later reprogrammed using C++, and a spatial component and visualization procedure were added. This spatial component was based on new research dealing with the effects of forest stand density on beetle movement. The density of trees in the pine forest greatly affects the population dynamics of the insect. Infestations of *D. frontalis* grow in size as a result of reproduction within individual trees. Brood beetles emerge from the tree and colonize adjacent trees, i.e., nearest neighbors. This behavior results in infestations having a directional growth pattern, an element of natural history unique to this insect. The density of trees in the forest, measured by foresters using a metric referred to as basal area (the cross-section of tree trunks at the breast height, expressed as ft^2/acre or m²/ha), greatly affects mortality of the insect as it searches for new hosts. This aspect of the biology of *D. frontalis* was incorporated into TAMBEETLE.

So a practical use of the TAMBEETLE model is in evaluating the effects of different silvicultural practices (planting strategies) on tree mortality resulting from herbivory by *D. frontalis*. Figure 10.8 illustrates the effect of thinning a forest to three different densities (70, 100, and 140 ft²/acre). One common approach to using spatial models in landscape ecology is to create different scenarios. *Scenarios* define the assumptions and parameters necessary to estimate potential future conditions, identify key processes, or reveal important interactions among simulated processes. In this example, TAMBEETLE was used to simulate population growth and tree mortality under two different stand conditions. The results are illustrated in Figure 10.9. The model clearly demonstrated that population

Figure 10.8 — Visualization of the basal area of pine forests (*Pinus* spp.) at three different stand densities: 70, 100, and 140 ft²/acre. Stand density influences southern pine beetle *(D. frontalis)* population dynamics in predictable ways (USDA, Forest Service photographs).

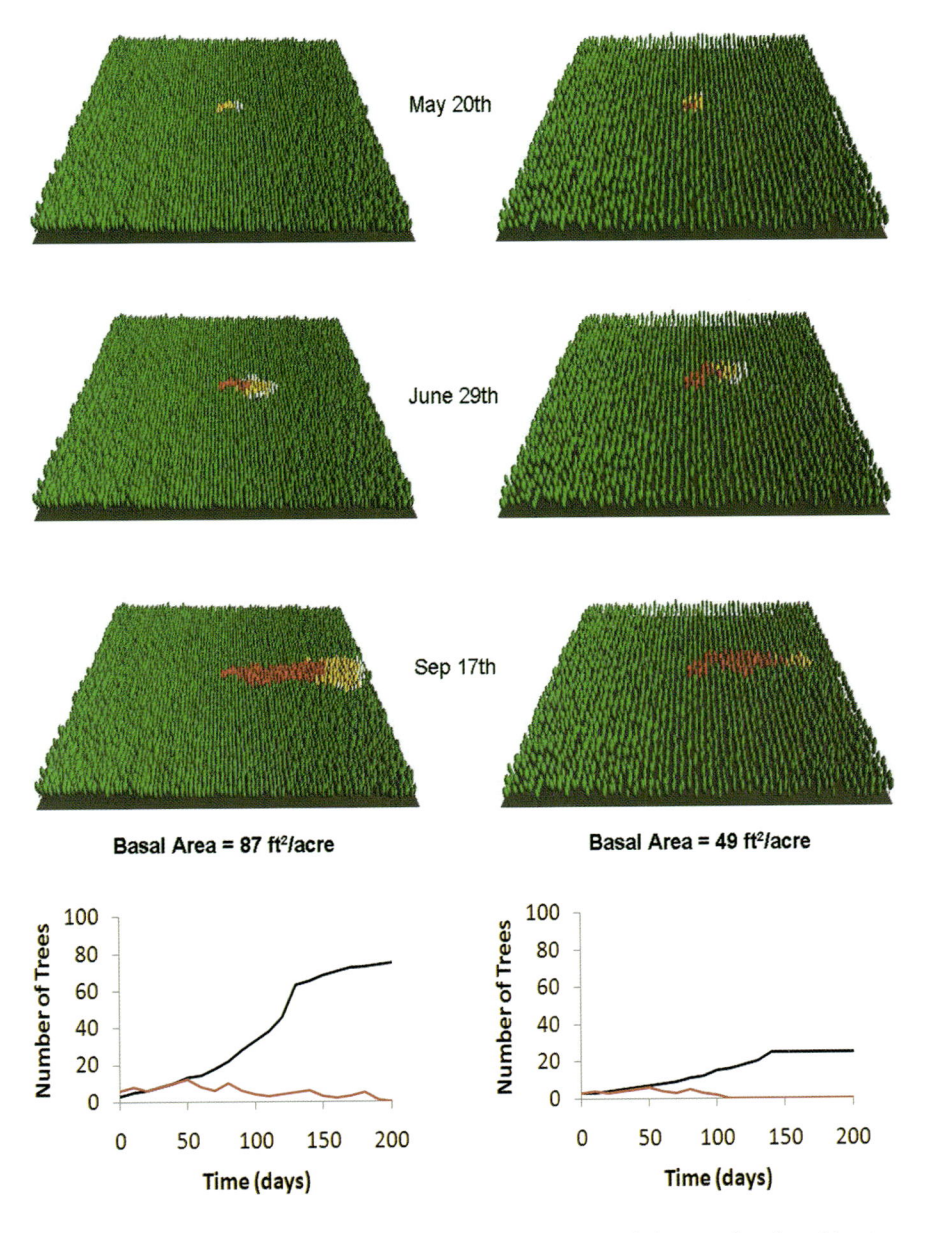

Figure 10.9 — Simulation of southern pine beetle (*D. frontalis*) population growth and resulting tree mortality under two scenarios of stand density (basal areas of 49 and 87 ft^2/acre) using a spatially explicit model (TAMBEETLE). The simulations demonstrated that population growth and tree mortality were affected by conditions of landscape structure, i.e., the density of forest trees in a component ecosystem. In the simulated forest: red = dead trees, yellow = currently infested trees, and white = trees under colonization. The black line representes the number of tress killed, and the red line is the number of currently infested trees (KEL image – illustration by Andrew Birt).

Landscape Analysis and Synthesis | 253

growth and the accompanying tree mortality were affected by conditions of landscape structure, i.e., the density of forest trees in a component ecosystem.

KNOWLEDGE ENGINEERING

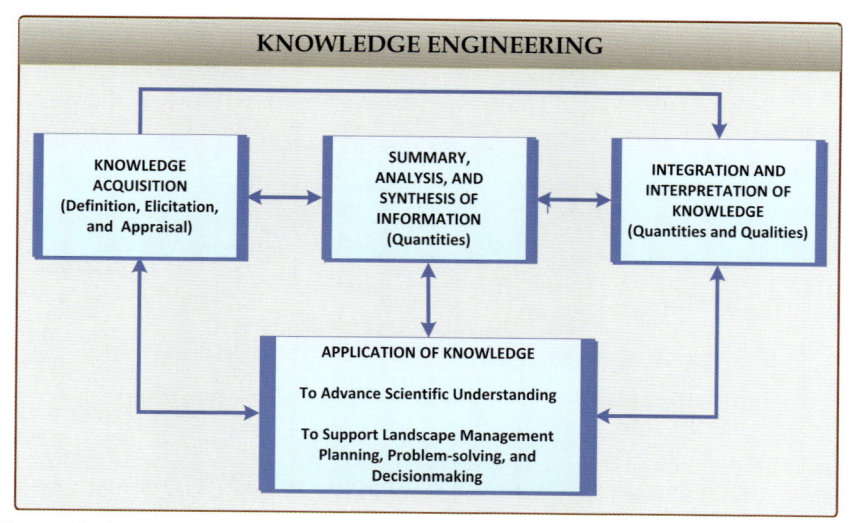

Knowledge engineering is an activity that embraces a set of concepts and methodologies dealing with (1) acquisition of knowledge, (2) analysis and synthesis of data and information [quantities], (3) integration and interpretation of knowledge [quantities and qualities], and (4) application of knowledge (Figure 10.10). In the context of landscape ecological investigations and landscape-use management, the goal of knowledge engineering is to facilitate use of the full extent of knowledge available on a subject of interest. In the case of landscape ecological investigations, interest is often directed to discovery of how ecological systems work or to contribute new knowledge to science. Ecologists have used personal flavors of knowledge engineering to define, conduct, summarize, and report discoveries from their research projects. In the case of landscape management, focus is directed to planning, designing, problem-solving, and decisionmaking.

Figure 10.10 — Knowledge engineering is an activity that embraces a set of concepts and methodologies dealing with (1) acquisition of knowledge, (2) analysis and synthesis of data and information [quantities], (3) integration and interpretation of knowledge [quantities and qualities], and (4) application of knowledge (KEL image).

254 | *Landscape Analysis and Synthesis*

In the preceding two sections of this chapter we examined basic concepts of landscape analysis and synthesis. Both of these activities involved the use of quantitative data. In the following section we redirect our focus to the issue of qualitative information in landscape ecology. Landscape-use management, planning, and design are activities that employ qualitative information and experiential knowledge, in addition to quantitative data. The methodologies associated with capturing qualitative information and heuristic knowledge differ from those used in scientific investigations. Two aspects of this component of knowledge engineering are examined: knowledge acquisition and knowledge integration.

Knowledge Acquisition

At the onset, all landscape ecological projects require an evaluation of the extant data and information that form the knowledge base for a specific problem of interest. Herein, we are using the term *project* in a generic sense to mean a landscape ecological research question or a landscape-use activity. There are three basic activities associated with the knowledge acquisition process: definition, elicitation, and appraisal. Each of these subjects is examined below.

Definition

The first planning task for all projects is an explicit problem statement, i.e., the specification of the fundamental or core problem that the project addresses. This statement cannot be ambiguous, and if multiple individuals are involved, there must be consensus among the participants. Once the problem is specified, *definition* involves the identification of relevant data and information. The approach involves a systematic evaluation of the problem of interest. Relevant numerical databases, maps, images, and printed materials can be located using the established Web-based, library, and networking approaches. Most landscape ecological projects will also require input from experts with different types of domain knowledge. So along with locating relevant data and information, it will be necessary to identify individuals who can provide the knowledge component needed for the project. The approaches to extracting knowledge from experts are different from those used in the identification of data and information. This topic is discussed in the next section.

Elicitation

Elicitation is the process and procedure for capturing experiential knowledge directly from experts and stakeholders. The goal is to develop a formal knowledge base for a particular topic or problem of interest that includes practical experience of experts. Two types of knowledge exist: explicit and tacit. *Explicit knowledge* can be expressed verbally and numerically and can be shared in the form of data, mathematical functions, text documents, etc. Explicit knowledge can be exchanged among individuals formally and systematically, and it can be processed by computer, transmitted electronically, or stored in databases. Explicit knowledge exchange can be facilitated using artificial intelligence-based techniques such as expert system and case-based reasoning systems (Desouza 2003). For example, through an interview process with a wildlife biologist who specializes in Bobwhite quail (*Colinus virginianus*), it would be possible assemble a set of rules that describe the behavior of this bird in a landscape ecological context. This explicit knowledge could then be used to develop an expert system for habitat management. By contrast, *tacit knowledge* is personal knowledge based on experiences, insights, intuition, observation, etc. It is influenced by intangible factors such as personal belief, perspective, and value system (Polanyi 1966). The often-cited example of tacit knowledge is riding a bicycle. Although this skill can be described in precise physics, explaining how it is done to someone who has not ridden a bicycle is problematic. The subjective and intuitive nature of tacit knowledge makes it difficult to process or transmit in a systematic manner. For tacit knowledge to be communicated it must be converted into words, models, or numbers (Desouza 2003). One way this issue is addressed in expert systems is through incorporation of fuzzy logic into rule bases. In the case of the Bobwhite quail example, rules dealing with habitat selection could be stated with probabilities associated with them, e.g., 70 percent of the time the bird will select a particular feature of habitat, but 30 percent of the time it will select another. Personal experience, judgment, and intuition of the wildlife biologists would guide the degree of fuzziness applied. Individuals involved directly in landscape-use management (foresters, farmers, ranchers, etc.) often develop immense tacit knowledge of the landscapes in their charge.

Experiential knowledge is particularly valuable in landscape ecological projects involving landscape-use and often is the principal representation available for planning, problem-solving, and decisionsupport. A variety of formal techniques have been devised to guide the knowledge elicitation process. Many of the techniques originated within psychology and they have become more structured, efficient, and systematic through computer-based implementation. See <http://www.epistemics.co.uk/> for a comprehensive review. Most of the knowledge elicitation procedures are flexible in their application and can be tailored to specific projects.

Appraisal

Appraisal deals with evaluation of the data and information that form the knowledge base for a particular problem. All scientists perform this task when they seek to place the results of their research into the existing corpus of knowledge. For complicated problems where a variety of sources and types of data and information are involved (e.g., evaluating the impact of global warming on biodiversity), knowledge-based systems are extremely useful for ordering and organizing extant data and information. These systems are also useful in identifying deficiencies in knowledge.

Knowledge Integration

Landscape ecology often deals with subjects that have large and disparate knowledge bases. The data and information that form the knowledge base for a specific problem often come from several different domain specialties, e.g., ecology, conservation biology, geography, sociology, economics, civil engineering, etc. The knowledge base can exist in several forms: (1) numerical data [usually stored in a database management system], (2) spatially referenced data themes [usually associated with a geographic information system], (3) numerical output from simulation models and mathematical evaluation functions, (4) text documents, and (5) heuristic knowledge of experts [based on human experience]. The knowledge base for most research questions in landscape ecology and problems in landscape-use management typically is a blend of these representations. Although the knowledge base for a topic of interest may be robust, it is also always incomplete and in a state of evolution.

The history and contemporary developments in knowledge integration are tied to business and engineering applications (Milton 2007, Milton 2008). In these domains, the principal product is usually an integrative knowledge-based system that serves to guide project management activities, manufacturing processes, engineering design, etc. The technologies associated with knowledge integration are particularly useful in landscape management.

Development of tools for integration, interpretation, and delivery of data, information and knowledge is an active area of the software engineering enterprise. Although the technological developments are beyond the interest of most landscape ecologists, the opportunities for application are manifold. An example of the application of integrative software for use in a landscape ecological application is the knowledge-based assessment of watershed condition reported by Reynolds et al. (2000). In this study the Ecosystem Management Decision Support System (EMDS) knowledge-based system was used. EMDS is an application framework for knowledge-based decision support of ecological landscape analysis at any geographic scale < http://www.institute.redlands.edu/emds/Default.aspx>.

EPILOGUE

So what conclusion can be made about landscape analysis and synthesis?

1. Landscape ecology is, in part, a computational science. The digital age facilitated and inspired the development of a myriad of computer-based methodologies for spatial data analysis, synthesis of information, and the integration of knowledge.

2. Landscape analysis deals with separating or breaking up of any whole into its parts so as to find out their nature, proportion, function, or relationship. In a landscape, the "whole" is represented as heterogeneity. In landscape ecology, we are interested in analysis of spatial data describing the constituent elements that form the landscape mosaic and in the measurements that define ecological phenomena, processes, and relationships taking place within the landscape environment.

3. In the context of landscape analysis, the definition of landscape heterogeneity (composition of parts of different kinds) was modified as

follows: the spatially structured variability of a property of interest, which may be a categorical or quantitative, explanatory or dependent variable.

4. Analysis in ecology and landscape ecology are different. In the ecological framework: patches/plots are internally homogeneous, plot context does not matter, and observations are spatially independent. In the landscape ecological framework: patches/plots may be internally heterogeneous, plot context may affect local processes, and observations may not be independent due to spatial interactions between local processes.

5. Two different approaches are used in the analysis of landscapes: landscape pattern indices (LPIs), which deal with data associated with categorical maps; and spatial statistics, which deal with point-data.

6. Landscape patch indices deal with four aspects of landscape structure: composition, configuration, shape, and scale (both spatial and temporal). Much of the need for spatial pattern indices is driven by interest in predicting the response of an ecological entity (e.g., fire, organisms, or nutrient flux) to the spatial heterogeneity of managed landscapes.

7. As with the data associated with the compositional components of the landscape, the quantitative measurements that characterize ecological phenomena, processes, and relationships are also spatially structured and vary through time. The measurements are represented as point locations in the landscape mosaic and can be associated the elements of the landscapes (e.g., habitat patches) or independent of a categorical classification. Point-data are collected by sampling.

8. Landscape synthesis deals with the putting together parts or elements so as to form a whole, i.e., it is the antithesis of analysis. In landscape ecology a great deal of emphasis has been placed on synthesis involving the use of spatially explicit simulation models, i.e., the entities associated with the model are defined by map coordinates and placed within a geographical context.

9. Landscape ecological research problems and landscape-use management projects involve the use of large and disparate knowledge bases. The goal of analysis and synthesis is to use the full extent of the knowledge base available about a subject of interest. Knowledge engineering is

the approach used to accomplish this end. Knowledge engineering is an activity that embraces a set of concepts and methodologies dealing with (1) acquisition of knowledge, (2) analysis and synthesis of data and information [quantities], (3) integration and interpretation of knowledge [quantities and qualities], and (4) application of knowledge.

Glossary

Aeolian process: In a landscape ecology context, the geomorphology process that deals with wind transport of entities.

Abiotic storage: The interplay of weathering (the breakdown of rock materials by mechanical and chemical means) and decomposition on soil-forming processes resulting in the assembly of pools of nutrients that can be used by primary producers for growth and development.

Allogenic ecosystem engineer: An organism that changes the landscape environment by transforming living or non-living materials from one physical state to another, via mechanical or other means, e.g., a beaver or an earthworm.

Analysis: Separating or breaking up of any whole into its parts so as to find out their nature, proportion, function, or relationship. In a landscape, the "whole" is represented as heterogeneity.

Attrition: The loss of elements in the landscape.

Autogenic ecosystem engineer: An organism that changes the landscape environment as a consequence of its inherent physical structures, i.e., their living or dead tissues, e.g., a forests or a reef.

Behavior: The response of an organism, group, or species to environmental factors.

Biodiversity: The variety and abundance of species that occur in an ecosystem or landscape of interest. For practical purposes biodiversity is synonymous with *species diversity*.

Biogeomorphology: The study of the interaction between ecology and geomorphology. The basic premise of this interaction is that the distribution and abundance of species is often related to underlying geomorphological landforms, while surface morphology may in turn be altered by living organisms.

Bioerosion: The process that deals with weathering and/or erosion of the land surface by organic means.

Bioconstruction: The process that deals with the production of sedimentary deposits, accretions, or accumulations by organic means, e.g., reef development.

Bioprotection: The process that deals with the roles of organisms in preventing or reducing the action or impact of other earth surface processes, e.g., the role of vegetation in controlling island bar development in fluvial systems.

Biosphere: That part of the earth and atmosphere capable of supporting living organisms.

Cartography: The art, science, and technology of making maps.

Cartograpic scale: The ratio between the size of an area on a map and the actual size of that area on the earth's surface.

Coastal processes: In geomorphology, the effects of waves, tides, and currents occurring at the interface between the terrestrial and marine (or other large water body) environments. The landscape ecology context centers on how waves, tides, and currents transport entities.

Commodity: Something useful or valuable to humans. In a landscape ecological context, commodities translate directly to ecosystem goods and services.

Community: Any assemblage of populations of living organisms in a prescribed area or habitat.

Condition: An abiotic environmental factor, which varies in space and time, to which living organisms are differentially responsive; the state of the environment, e.g., temperature, humidity, wind speed, etc.

Consequence: The effect(s) resulting from the propagation of an entity from one place to another in a landscape. Examples of consequences are soil erosion, disease epidemics, and pollination of flowering plants.

Consumption: The total intake of food or energy by an organism during a specified period of time.

Corridor: A narrow strip of land (or water) that differs from the areas adjacent to it on both sides. The adjacent areas can be the matrix, patches, or another type of corridor.

Data: The measurements that define an ecological phenomenon, process, or relationship of interest.

Decomposition: The breakdown of complex energy-rich organic molecules into simple inorganic constituents.

Dispersal: Movement of an individual or a population away from its place of birth or origin.

Dissection: Separating or cutting a landscape into pieces.

Disturbance: An initiating cause (physical force, a process, or an event) that produces an effect (consequence) that is greater than average, normal, or expected. This definition requires a reference state, i.e., a mean condition bounded by a range in variation.

Disturbance impact: Any effect on the landscape environment resulting from a disturbance event.

Social (axiological) *impact* refers to the effects of disturbance on aesthetic, moral, and metaphysical values associated with the landscape.

Political impact refers to the effects of disturbances on the landscape environment that result in actions, practices, policies of local, state, or federal governmental agencies.

Economic impact is simply defined as the effect of disturbances on the monetary receipts from the production of goods and services from a landscape.

Ecological impact is the qualitative or quantitative change in conditions and/or resources in the landscape environment resulting from a disturbance event.

Disturbance regime: The ensemble of disturbance types associated with a specific landscape environment.

Driver: A human-initiated cause (physical force, a process, or an event) that results in landscape-use change. This definition is restrictive to landscape-use change as there are also drivers of landscape-cover change that are not human-initiated.

Ecology: The study of how organisms interact with their environment.

Ecoregion (=*biome*): A biogeographical region or formation.

Ecosystem: The biotic community plus its abiotic environment.

Ecosystem cluster (=*geocomplex, meta-ecosystem*): A group of ecosystems (landscape elements) connected by significant exchange of energy, materials, and information.

Ecosystem engineer: An organism that mediates changes to the abiotic environment through its influence on the resources and conditions that define habitats for the community of life that forms the biotic component of the ecosystem.

Ecosystem management: The orchestrated modification or manipulation of the basic ecosystem processes (primary production, consumption, decomposition, and abiotic storage) for desired human-defined ends.

Ecotone: A transition area occurring at the interface of two or more distinct landscape elements.

Ecotope: A bounded ecosystem.

Edge effects (=*edge influences*): Changes in the resources and conditions of the ecotone that result from the adjacency of the landscape elements. The edges create an ecotone environment that is different from that of the adjacent landscape elements.

Edge species: A species adapted to edge habitat, i.e., that occurs in ecotones separating adjacent landscape elements.

Elicitation: The process and procedure for capturing experiential knowledge directly from experts and stakeholders.

Emergent properties: Properties of the whole not reducible to the sum of the properties of the parts.

Entity: Something that has separate and distinct existence and objective or conceptual reality. Entities represent what is being propagated or transported within the landscape. Entities can be classed as matter, information, and energy. The entities include many specific forms, e.g., dust particle, pollen grain, water, herbivores, etc.

Explicit knowledge: A representation of knowledge that can be expressed verbally and numerically and can be shared in the form of data, mathematical functions, text documents, etc.

Fitness: The relative contribution that an individual makes to the gene pool of the next generation. The fittest individuals in a population are those that leave the greatest number of descendants, relative to the number of descendents left by other individuals in a population.

Fluvial Process: In a landscape ecology context, the geomorphology process that deals with water transport of entities.

Fragmentation: Breaking up a landscape into smaller pieces.

Functional connectivity: A landscape environment where requisite habitat patches are present, sufficient in size, and accessible for an individual species.

Functional heterogeneity: How an organism perceives and responds to the landscape environment.

Geographic range: The spatial extent of all home ranges of a population or species.

Geomorphology: The science of landforms, including their history and origin.

Glacial movement: In a landscape ecological context, the movement of entities associated with a glacier.

Hard boundary: An ecotone produced by high contrast edges, e.g., the intersection of a forest patch and a grassland patch.

Habitat: The physical place where an organism either actually or potentially lives. Habitat is described using the biotic and abiotic features of the environment that are thought be important to an organism.

Habitat connectivity: The connectedness between patches of suitable habitat in the landscape. This attribute of the landscape environment is defined by maintenance movement behavior and dispersal capabilities of the species in relation to the spatial locations of the habitat patches.

Habitat heterogeneity: The different places in a landscape suitable for occupancy and use by a living organism, i.e., the variety of physical places where an organism either actually or potentially lives. These habitat places can include patches, corridors, matrix, and ecotones.

Heterogeneity: Composition of parts of different kinds; the spatially structured variability of a property of interest, which may be a categorical or quantitative, explanatory or dependent variable.

Hierarchy: A series of consecutively subordinate categories forming a system of classification.

Home range: The area where the life history of an animal is played out. This area provides the resources and conditions needed for survival, growth, and reproduction and is defined in extent by the behavioral motivational states of the animal.

Impact: Any effect on the landscape environment resulting from a disturbance event.

Information: Data that have been given meaning by way of relational connection.

Interior species: Species adapted to interior habitats, i.e., the central portion of a landscape element such as a patch.

Invasive species: A non-indigenous (non-native) organism that has been introduced into a novel ecosystem, i.e., an ecosystem within a landscape that occurs outside the natural range or potential dispersal distance of the species.

Initiating cause: A factor that brings about an effect or a result, e.g., a fire, hurricane, herbivory, introduction of a pathogen, etc.

Keystone species: A species whose effect on an ecosystem is disproportionately large relative to its low biomass in the community as a whole.

Knowledge: Contextually integrated information, i.e., knowledge consists of an organized body of information.

Knowledge engineering: An activity that embraces a set of concepts and methodologies dealing with acquisition of knowledge, analysis and synthesis of data and information [quantities], integration and interpretation of knowledge [quantities and qualities], and the application of knowledge.

Landform (=physiography): The configuration of the land surface having distinctive character and produced by natural processes. A landform includes surface geometry and underlying geologic material.

Landscape: A spatially explicit geographic area, i.e., an area defined by coordinates, consisting of recognizable and characteristic component entities. The entities are variously referred to in the literature as ecosystems, ecotopes, sites, elements, tessera, units, etc.

Landscape change: The alteration of structure and function of the landscape environment over time.

Landscape-cover change: Alteration of biophysical attributes of the landscape environment resulting from biogeomorphology, the activities of living organisms (other than humans), and natural disturbances.

Landscape configuration: The spatial arrangement (juxtaposition) of the basic landscape elements (ecosystems) in relation to one another. Emphasis can be directed to individual ecosystems or to clusters (neighborhoods).

Landscape composition: The number and relative frequency of a landscape element (ecosystem).

Landscape-use change: Alteration of biophysical attributes of the landscape environment resulting from human purpose or intent.

Landscape domestication: The activities of humans that structurally shape and functionally modify landscapes to satisfy basic human needs.

Landscape ecology: The science that embraces the agenda of ecology (however broadly or narrowly defined) in a spatially explicit manner. An ecological question is a landscape ecology question if the answer requires consideration of spatial components.

Landscape function: The flux of energy, information, and materials within and among the component ecosystems (elements) forming the landscape.

Landscape environment: All the external conditions and resources (living and nonliving) that affect an organism or other specified system (e.g., a forest landscape) during its life time.

Landscape mosaic: The pattern formed by elements (ecosystems) that make up a landscape.

Landscape structure: The components of the landscape and their linkages and configurations.

Landscape transformation: Change in form, appearance, nature, or character of the landscape.

Level of integration: A compartment containing specific information, concepts, and principles that collectively represent the knowledge base for a subject of interest.

Life cycle: For a species, the sequence of morphological stages and physiological processes that link one generation to the next.

Life history: The significant features of the life cycle through which a species passes, with particular references to strategies influencing survival, growth, and reproduction.

Maintenance movement: Animal movement resulting from behavioral motivational states, i.e., movement occurring in association with behavior involved in searching for food, water, mates, refuges; parental care; predator avoidance; playing; etc.

Map: A graphic representation of the cultural and physical environment.

Mass movement: In a landscape ecology context, the geomorphology process that deals with a gravity-driven downward movement of slope material without the assistance of moving water, ice, or air.

Materials: The elements, constituents, or substances of which something is composed or can be made.

Matrix: The background or cover type of a landscape, i.e., the backdrop into which patches and corridors are imbedded. The basic attributes used to delineate the matrix in a landscape include area, connectivity, and control over dynamics.

Metrology: The field of study that addresses measurement.

Migration: The seasonal movement of a population from one geographical location to another and back again. It is usually bi-annual or seasonal and generally implies breeding site philopatry.

Movement (=displacement): A change in place or position.

Niche: A multidimensional space within which the environment (resources and conditions) permits an individual species to survive, grow, reproduce, and maintain a viable population (persist). For an individual species, the dimensions of this space are defined as a function of the inherent tolerances to environmental conditions and resource requirements as well as interactions with other organisms.

Nutrient cycling: The transformation of chemical elements from inorganic form in the environment to organic form in organisms, and via decomposition back into inorganic form.

> *Organismal phase*: The nutrients exist in organic form as the living tissues of organisms.

> *Environmental phase*: The chemical nutrients are in inorganic form and occur in the soil, water, or air.

Patch: A surface area differing from its surroundings in nature or appearance; a bounded area (i.e., an area that can be defined by coordinates) embedded in a matrix (or other landscape element). Synonyms include piece, scrap, and bit. Patches are a fundamental structural element in most landscapes.

Perforation: The making of holes in a landscape.

Primary production: The conversion of solar energy by green plants (via photosynthesis) to organic substances.

Physical ecosystem engineering: Change to the abiotic environment caused by the activities of living organisms that results in the creation, modification, maintenance, or destruction of habitats, i.e., structurally mediated modification of habitats by organisms.

Population: A collective group of individuals of one species that inhabits an area sufficiently small to enable interbreeding and functions as part of a biotic community.

Process: A change that leads towards a predictable result.

Propagation vector: The means for transport or conveyance of the entity, e.g., wind, water, gravity, animal locomotion, etc.

Resources: All things consumed by an organism. Resources are quantities that can be reduced by the activity of living organisms.

Road ecology: The branch of ecology that explores and addresses the relationship between the natural environment and the road system.

Scalar: In mathematics, a quantity that can be fully described by a magnitude or numerical value alone, e.g., energy.

Shrinkage: Decreasing the size of selected elements within the landscape.

Sink: An end point, an area, or a reservoir where input of an entity exceeds the output of it. Entities refer specifically to energy, materials, and information.

Soft boundary: An ecotone produced by low contrast edges, e.g., the intersection of a prairie grassland with a coastal bermuda pasture.

Source: A starting point, an area, or a reservoir where output of an entity exceeds the input of it. Entities refer specifically to energy, materials, and information.

Spatial extent (=*scale*): The size of an area of interest. Size is usually expressed as m, m^2, or m^3.

Spatially explicit: A concept in which the entities associated with the landscape are defined by map coordinates and placed within a geographical context.

Stepping stone: A patch that serves as a means of progress or advancement, usually in reference to movement of living organisms in the landscape.

Synthesis: The putting together of parts or elements so as to form a whole, i.e., it is the antithesis of analysis.

Tacit knowledge: Personal knowledge based on experiences, insights, intuition, observation, etc. It is influenced by intangible factors such as personal belief, perspective, and value system.

Temporal extent (*=scale*): The length of time for an observation or an event to take place. Time is usually expressed in chronological units (seconds, hours, days, months, years) or occasionally in relation to events (daily, lunar, annual cycles).

Weathering: The adjustment of the chemical, mineralogical, and physical properties of rocks in response to environmental conditions prevailing at the earth's surface.

Chemical weathering leads to the release of compounds in solution with the creation of new mineral products.

Physical weathering results in the breakdown of original rock material into smaller particles.

References

Ahern, J. 2005. Integration of landscape ecology and landscape architecture: An evolutionalry and reciprocal process. In J. A. Wiens and M. R. Moss (eds.), *Issues and Perspectives in Landscape Ecology*. Cambridge University Press, Cambridge.

Ahl, V. and T. F. H. Allen. 1996. *Hierarchy Theory: A Vision, Vocabulary, and Epistemology*. Columbia University Press, New York.

Allee, W. C., A. E. Emerson, O. Park, T. Park, and K. P. Schmidt. 1949. *Principles of Animal Ecology*. W.B. Saunders Company, Philadelphia.

Allen, T. F. H. and T. B. Starr. 1982. *Hierarchy: Perspectives for Ecological Complexity*. University of Chicago Press, Chicago.

Alpers, S. 1983. *The Art of Describing: Dutch Art in the Seventeenth Century*. University of Chicago Press, Chicago.

Anderson, J. R., E. E. Hardy, J. T. Roach, and R.E. Witmer. 1976. *A Land-Use Classification System for Use with Remote Sensor Data*. Geological Survey Paper 964. U.S. Geological Survey, Washington, D.C.

Antrop, M. 2005. Why landscapes of the past are important for the future? *Landscape and Urban Planning* 70: 21-34.

Bailey, R. G. 1996. *Ecosystem Geography*. Springer, New York.

Bailey, R. G. 1998. *Ecoregions: The Ecosystem Geography of the Oceans and Continents*. Springer, New York.

Bailey, R. G. 2002. *Ecoregion-Based Design for Sustainability*. Springer, New York.

Baum, K., K. Haynes, F. Dillemuth, and J. Cronin. 2004. The matrix enhances the effectiveness of corridors and stepping stones. *Ecology* 85: 2671-2676.

Baum, K., W. Rubink, M. Pinto, and R. Coulson. 2005. Spatial and temporal distribution and nest site characteristics of feral honey bee (Hymenoptera: Apidae) colonies in a coastal prairie landscape. *Environmental Entomology* 34: 610-618.

Baum, K., M. Tchakerian, S. Thoenes, and R. Coulson. 2008. Africanized honey bees in urban environments: A spatio-temporal analysis. *Landscape and Urban Planning* 85: 123-132.

Begon, M., C. Townsend, and J. Harper. 2006. *Ecology: From Individuals to Ecosystems*. Blackwell Publishing, Malden, MA.

Bender, D. and L. Fahrig. 2005. Matrix structure obscures the relationship between interpatch movement and patch size and isolation. *Ecology* 86: 1023-1033.

Bisson, P. A., C. M. Crisafulli, B. R. Fransen, R. E. Lucas, and C. P. Hawkins. 2005. Responses of fish to the 1980 eruption of Mount St. Helens. In V. H. Dale, F. J. Swanson, and C. M. Crisafulli (eds.), *Ecological Response to the 1980 Eruptions of Mount St. Helens*. Springer, New York.

Bissonette, J. A. and I. Storch. 2003. *Landscape Ecology and Resource Management: Linking Theory with Practice*. Island Press, Washington, D.C.

Bonner, J. T. 1993. *Life Cycles: Reflections of an Evolutionary Biologist*. Princeton University Press, Princeton, NJ.

Bormann, F. H. and G. E. Likens. 1979. *Pattern and Process in a Forested Ecosystem: Disturbance, Development, and the Steady State Based on the Hubbard Brook Ecosystem Study.* Springer-Verlag, New York.

Boyce, S. G. 1995. *Landscape Forestry.* Wiley, New York.

Boyd, J. and S. Banzhaf. 2007. What are ecosystem services? The need for standardized environmental accounting units. *Ecological Economics* 63: 616-626.

Buchmann, S. L. and G. P. Nabhan. 1996. *The Forgotten Pollinators.* Island Press/Shearwater Books, Washington, D.C.

Buckley, Y. 2008. The role of research for integrated management of invasive species, invaded landscapes, and communities. *Journal of Applied Ecology* 45: 397-402.

Bürgi, M., A. Hersperger, and N. Schneeberger. 2004. Driving forces of landscape change — current and new directions. *Landscape Ecology* 19: 857-868.

Burrough, P. A. 1986. *Principles of Geographical Information Systems for Land Resource Assesment.* Oxford University Press, Oxford.

Butler, D. R. 1995. *Zoogeomorphology: Animals as Geomorphic Agents.* Cambridge University Press, Cambridge.

Cairns, D. M., C. W. Lafon, J. D. Waldron, M. D. Tchakerian, R. N. Coulson, K. D. Klepzig, A. G. Birt, and W. Xi. 2008. Simulating the reciprocal interaction of forest landscape structure and southern pine beetle herbivory using LANDIS. *Landscape Ecology* 23: 403-415.

Campbell, N. and J. Reece. 2005. *Biology.* Benjamin-Cummings Publishing Company, San Francisco.

Carson, R. 1962. *Silent Spring.* Houghton Mifflin, Boston.

Casey, E. S. 2002. *Representing Place: Landscape Painting and Maps.* University of Minnesota Press, Minneapolis.

Chapin, F. S., P. A. Matson, and H. A. Mooney. 2002. *Principles of Terrestrial Ecosystem Ecology*. Springer, New York.

Clark, K. 1976. *Landscape into Art*. Beacon Press, Boston.

Clark, W. C. 1985. Scales of climate impacts. *Climatic Change* 7: 5-27.

Coleman, D. C., D. A. Crossley Jr., and P. F. Hendrix. 1996. *Fundamentals of Soil Ecology*. Elsevier Academic Press, Burlington, MA.

Coleman, D. C. and P. F. Hendrix. 2000. *Invertebrates as Webmasters in Ecosystems*. CABI Pub., Wallingford, Oxon, UK.

Corner, J. and A. S. MacLean. 1996. *Taking Measures: Across the American Landscape*. Yale University Press, New Haven, CT.

Costanza, R., R. d'Arge, R. de Groot, S. Farber, M. Grasso, B. Hannon, K. Limburg, S. Naeem, R. O'Neill, J. Paruelo, R. Raskin, P. Sutton, and M. van den Belt. 1997. The value of the world's ecosystem services and natural capital. *Nature* 387: 253-260.

Coulson, R. N., R. M. Feldman, P. J. H. Sharpe, P. E. Pulley, T. L. Wagner, and T. L. Payne. 1989. An overview of the TAMBEETLE model of *Dendroctonus frontalis* population dynamics. *Ecography* 12: 445-450.

Coulson, R. N., L. Folse, and D. K. Loh. 1987. Artificial-intelligence and natural resource management. *Science* 237: 262-267.

Coulson, R. N., M. D. Guzman, K. Skordinski, J. W. Fitzgerald, R. N. Conner, D. C. Rudolph, F. L. Oliveria, D. F. Wunneburger, and P. E. Pulley. 1999. Forest landscapes: Their effect on the interaction of the southern pine beetle and Red-cockaded woodpecker. *Journal of Forestry* 97: 4-11.

Coulson, R. N., M. A. Pinto, M. D. Tchakerian, K. A. Baum, W. L. Rubink, and J. S. Johnston. 2005. Feral honey bees in pine forest landscapes of east Texas. *Forest Ecology and Management* 215:91-102.

Coulson, R. N. and F. M. Stephen. 2006. Impacts of insects in forest landscapes: Implications for forest health management. In T. D. Paine (ed.), *Invasive Forest Insects, Introduced Forest Trees, and Altered Ecosystems*. Springer, New York.

Coulson, R. N. and J. A. Witter. 1984. *Forest Entomology: Ecology and Management*. Wiley, New York.

Coulson, R. N. and D. F. Wunneburger. 2000. Impacts of insects on human-dominated and natural forest landscapes. In D. C. Coleman and P. F. Hendrix (eds.), *Invertebrates as Webmasters in Ecosystems*. CABI Pub., Wallingford, Oxon, UK.

Crooks, J. 2002. Characterizing ecosystem-level consequences of biological invasions: The role of ecosystem engineers. *Oikos* 97: 153-166.

Crossley D. A., Jr., G. J. House, R. M. Snider, R. J. Snider, and B. R. Stinner. 1984. The positive interactions in agroecosystems. In R. Lowrance, B. R. Stinner, and G. J. House (eds.), *Agricultural Ecosystems: Unifying Concepts*. Wiley, New York.

Daily, G. C. 1997. *Nature's Services: Societal Dependence on Natural Ecosystems*. Island Press, Washington, D.C.

Dale, V., S. Brown, R. Haeuber, N. Hobbs, N. Huntly, R. Naiman, W. Riebsame, M. Turner, and T. Valone. 2000. Ecological principles and guidelines for managing the use of land. *Ecological Applications* 10: 639-670.

Davic, R. 2003. Linking keystone species and functional groups: A new operational definition of the keystone species concept - response. *Conservation Ecology* 7: Response 11.

de Groot, R. S., M. A. Wilson, and R. M. J. Boumans. 2002. A typology for the classification, description and valuation of ecosystem functions, goods, and services. *Ecological Economics* 41: 393-408.

Delcourt, P. A. and H. R. Delcourt. 1992. Ecotone dynamics in space and time. In A. J. Hansen and F. Di Castri (eds.), *Landscape Boundaries: Consequences for Biotic Diversity and Landscape Flows*. Springer-Verlag, New York.

Desouza, K. 2003. Facilitating tacit knowledge exchange. *Communications of the ACM* 46: 85-88.

Despommier, D., B. Ellis, and B. Wilcox. 2007. The role of ecotones in emerging infectious diseases. *Ecohealth* 3: 281-289.

Didham, R., J. Tylianakis, M. Hutchison, R. Ewers, and N. Gemmell. 2005. Are invasive species the drivers of ecological change? *Trends in Ecology & Evolution* 20: 470-474.

Donald, P. F. and A. D. Evans. 2006. Habitat connectivity and matrix restoration: The wider implications of agri-environmental schemes. *Journal of Applied Ecology* 43: 209-218.

Fahrig, L. 2002. Effect of habitat fragmentation on the extinction threshold: A synthesis. *Ecological Applications* 12: 346-353.

Farina, A. 2000a. *Landscape Ecology in Action*. Kluwer Academic, Dordrecht, The Netherlands.

Farina, A. 2000b. *Principles and Methods in Landscape Ecology: Toward a Science of Landscape*. Kluwer Academic, Dordrecht, The Netherlands.

Feldman, R. M., G. L. Curry, and R. Coulson. 1981. Mathematical model of field population dynamics of the southern pine beetle *Dendroctonus frontalis*. *Ecological Modelling* 13: 261-282.

Fischer, J. and D. Lindenmayer. 2007. Landscape modification and habitat fragmentation: A synthesis. *Global Ecology and Biogeography* 16: 265-280.

Fisher, B., K. Turner, M. Zylstra, R. Brower, R. de Groot, S. Farber, P. Ferraro, R. Green, D. Hadley, J. Harlow, P. Jefferiss, C. Kirkby, P. Morling, S. Mowatt, R. Naidoo, J. Paavola, B. Strassburg, D. Yu, and A. Balmford. 2008. Ecosystem services and economic theory: Integration for policy-relevant research. *Ecological Applications* 18: 2050-2067.

Flather, C. H. and M. Bevers. 2002. Patchy reaction-diffusion and population abundance: The relative importance of habitat amount and arrangement. *American Naturalist* 159:40-56.

Forman, R. T. T. 1995. *Land Mosaics: The Ecology of Landscapes and Regions.* Cambridge University Press, Cambridge.

Forman, R. T. T. and M. Godron. 1981. Patches and structural components for a landscape ecology. *BioScience* 31: 733-740.

Forman, R. T. T. and M. Godron. 1986. *Landscape Ecology.* Wiley, New York.

Forman, R. T., R. T. Sperling, J. A. Bissonette, A. P. Clevenger, C. D. Cutshall, V. H. Dale, L. Fahrig, R. France, C. R. Goldman, K. Heanue, J. A. Jones, F. J. Swanson, T. Turrentine, and T. C. Winter. 2003. *Road Ecology: Science and Solutions.* Island Press, Washington, D.C.

Fortin, M-J. and M. Dale. 2005. *Spatial Analysis: A Guide for Ecologists.* Cambridge University Press, Cambridge.

Fritts, T. and G. Rodda. 1998. The role of introduced species in the degradation of island ecosystems: A case history of Guam. *Annual Review of Ecology and Systematics* 29: 113-140.

Gaucherel, C. and T. Houet. 2009. Preface to the selected papers on spatially explicit landscape modelling: Current practices and challenges. *Ecological Modelling* 220: 3477-3480.

Gergel, S. E. and M. G. Turner. 2002. *Learning Landscape Ecology: A Practical Guide to Concepts and Techniques.* Springer, New York.

Gersmehl, P. 1996. *The Language of Maps.* National Council for Geographic Education, Indiana, PA.

Gibson, J. J. 1986. *The Ecological Approach to Visual Perception.* Houghton Mifflin, Boston.

Golley, F. B. 1993. *A History of the Ecosystem Concept in Ecology: More than the Sum of the Parts.* Yale University Press, New Haven, CT.

Gregory, S. V., F. J. Swanson, W. A. McKee, and K. W. Cummins. 1991. An ecosystem perspective of riparian zones: Focus on links between land and water. *BioScience* 41: 540-549.

Gunderson, L. H., C. S. Holling, and S. S. Light. 1995. *Barriers and Bridges to the Renewal of Ecosystems and Institutions.* Columbia University Press, New York.

Gunderson, L. H. and C. S. Holling. 2002. *Panarchy: Understanding Transformations in Human and Natural Systems.* Island Press, Washington, D.C.

Gustafson, E. J. 1998. Minireview: Quantifying landscape spatial pattern: What is the state of the art? *Ecosystems* 1: 143-156.

Gustafson, E.J. 2002. Simulating changes in landscape pattern. In S.E. Gergel and M. G. Turner (eds.), *Learning Landscape Ecology.* Springer, New York.

Gustafson, E. J. 2006. What is landscape ecology? <http://www.usiale.org>.

Gutzwiller, K. J. 2002. *Applying Landscape Ecology in Biological Conservation.* Springer, New York.

Haila, Y. 2002. A conceptual genealogy of fragmentation research: From island biogeography to landscape ecology. *Ecological Applications* 12: 321-334.

Haines-Young, R. 2005. Landscape pattern: Context and process. In J. A. Wiens and M. R. Moss (eds.), *Issues and Perspectives in Landscape Ecology.* Cambridge University Press, Cambridge.

Hanes, T. L. 1971. Succession after fire in chaparral of Southern California. *Ecological Monographs* 41: 27-52.

Hansen, A. J., F. Di Castri, and A. D. Armand. 1992. *Landscape Boundaries: Consequences for Biotic Diversity and Ecological Flows.* Springer-Verlag, New York.

Hansson, L., L. Fahrig, and G. Merriam. 1995. *Mosaic Landscapes and Ecological Processes.* Chapman & Hall, London.

Hargrove, W. and F. Hoffman. 2004. Potential of multivariate quantitative methods for delineation and visualization of ecoregions. *Environmental Management* 34: S39-S60.

Harper, K., S. Macdonald, P. Burton, J. Chen, K. Brosofske, S. Saunders, E. Euskirchen, D. Roberts, M. Jaiteh, and P. Esseen. 2005. Edge influence on forest structure and composition in fragmented landscapes. *Conservation Biology* 19: 768-782.

He, H. 2008. Forest landscape models: Definitions, characterization, and classification. *Forest Ecology and Management* 254: 484-498.

Hersperger, A. 2006. Spatial adjacencies and interactions: Neighborhood mosaics for landscape ecological planning. *Landscape and Urban Planning* 77: 227-239.

Hess, G. R. and R. A. Fischer. 2001. Communicating clearly about conservation corridors. *Landscape and Urban Planning* 55: 195-208.

Hobbs, R. J. and L. F. Huenneke. 1992. Disturbance, diversity, and invasion: Implications for conservation. *Conservation Biology* 6: 324-337.

Holland, M., P. G. Risser, and R. J. Naiman. 1991. *Ecotones: The Role of Landscape Boundaries in the Management and Restoration of Changing Environments.* Chapman and Hall, New York.

Holling, C. S. 1986. Resilience of ecosystems: Local surprise and global change. In W. C. Clark and R. E. Munn (eds.), *Sustainable Development of the Biosphere.* Cambridge University Press, Cambridge.

Holling, C. S. 1992. Cross-scale morphology, geometry, and dynamics of ecosystems. *Ecological Monographs* 62: 447-502.

Holling, C. S. 2001. Understanding the complexity of economic, ecological, and social systems. *Ecosystems* 4: 390-405.

Holling, C. S. 2004. From complex regions to complex worlds. *Ecology and Society* 9: Article 11.

Hooper, D., F. Chapin, J. Ewel, A. Hector, P. Inchausti, S. Lavorel, J. Lawton, D. Lodge, M. Loreau, S. Naeem, B. Schmid, H. Setala, A. Symstad, J. Vandermeer, and D. Wardle. 2005. Effects of biodiversity on ecosystem functioning: A consensus of current knowledge. *Ecological Monographs* 75: 3-35.

Hoyle, Z. 2008. Red-cockaded woodpeckers and hurricanes. *Compass* 12: 12-13.

Hutchinson, G. E. 1957. Concluding remarks. *Cold Springs Harbor Symposia on Quatitative Biology.*

Isard, S. A. and S. H. Gage. 2001. *Flow of Life in the Atmosphere: An Airscape Approach to Understanding Invasive Organisms.* Michigan State University Press, East Lansing, MI.

Johnston, J. C. 1995. Effects of animals on landscape pattern. In L. Hansson, L. Farhig, and G. Merriam (eds.), *Mosaic Landscapes and Ecological Processes.* Chapman and Hall, London.

Jones, C., J. Lawton, and M. Shachak. 1994. Organisms as ecosystem engineers. *Oikos* 69: 373-386.

Jones, C., J. Lawton, and M. Shachak. 1997. Positive and negative effects of organisms as physical ecosystem engineers. *Ecology* 78: 1946-1957.

Kareiva, P., S. Watts, R. McDonald, and T. Boucher. 2007. Domesticated nature: Shaping landscapes and ecosystems for human welfare. *Science* 316: 1866-1869.

Karr, J. R. 2002. What from ecology is relevant to design and planning? In B. Johnson and K. Hill (eds.), *Ecology and Design: Frameworks for Learning.* Island Press, Washington, D.C.

Kearney, M. 2006. Habitat, environment, and niche: What are we modelling? *Oikos* 115: 186-191.

Keet, C. M. 2006. Representations of the ecological niche. In B. Klein, I. Johansson, and T. Roth-Berghofer (eds.), *Third international workshop on philosophy and informatics* (wspi2006), Saarbrucken, Germany. 3-4 May 2006. *IFOMIS Reports.*

Kelmelis, J. A. 1998. Process dynamics, temporal extent, and causal propagation as the basis for linking space and time. In M. J. Egenhofer and R. G. Golledge (eds.), *Spatial and Temporal Reasoning in Geographic Information Systems.* Oxford Univeristy Press, Oxford.

Klopatek, J. M. and R. H. Gardner. 1999. *Landscape Ecological Analysis: Issues and Applications.* Springer, New York.

Kolasa, J., S. T. Pickett, and T. F. H. Allen. 1991. *Ecological Heterogeneity.* Springer-Verlag, New York.

Kolasa, J. and C. D. Rollo. 1991. Introduction: The heterogeneity of heterogeneity — a glossary. In J. Kolasa, S. T. Pickett, and T. F. H. Allen (eds.), *Ecological Heterogeneity.* Springer-Verlag, New York.

Lambin, E., B. Turner, H. Geist, S. Agbola, A. Angelsen, J. Bruce, O. Coomes, R. Dirzo, G. Fischer, C. Folke, P. George, K. Homewood, J. Imbernon, R. Leemans, X. Li, E. Moran, M. Mortimore, P. Ramakrishnan, J. Richards, H. Skanes, W. Steffen, G. Stone, U. Svedin, T. Veldkamp, C. Vogel, and J. Xu. 2001. The causes of land-use and land-cover change: Moving beyond the myths. *Global Environmental Change-Human and Policy Dimensions* 11: 261-269.

Landres, P., P. Morgan, and F. Swanson. 1999. Overview of the use of natural variability concepts in managing ecological systems. *Ecological Applications* 9: 1179-1188.

Lawton, J. 1994. What do species do in ecosystems? *Oikos* 71: 367-374.

Levin, S. A. 1992. The problem of pattern and scale in ecology: The Robert H. MacArthur Award Lecture. *Ecology* 73: 1943-1967.

Lewontin, R. C. 2000. *The Triple Helix: Gene, Organism, and Environment.* Harvard University Press, Cambridge, MA.

Li, H. and J. Wu. 2004. Use and misuse of landscape indices. *Landscape Ecology* 19: 389-399.

Liebhold, A. and J. Gurevitch. 2002. Integrating the statistical analysis of spatial data in ecology. *Ecography* 25: 553-557.

Likens, G. 1992. An Ecosystem Approach: Its Use and Abuse. *Excellence in Ecology, Book 3.* Ecology Institute, Oldendorf/Luhe, Germany.

Liu, J. and W. W. Taylor. 2002. *Integrating Landscape Ecology into Natural Resource Management.* Cambridge University Press, Cambridge.

Lockwood, J., M. Hoopes, and M. Marchetti. 2007. *Invasion Ecology.* Wiley-Blackwell, Malden, MA.

Loreau, M., N. Mouquet, and R. Holt. 2003. Meta-ecosystems: A theoretical framework for a spatial ecosystem ecology. *Ecology Letters* 6: 673-679.

Lovelady, C. N., P. E. Pulley, R. N. Coulson, and R. O. Flamm. 1991. Relation of lightning to herbivory by the southern pine bark beetle guild (Coleoptera: Scolytidae). *Environmental Entomology* 20: 1279-1284.

Lubchenco, J., A. M. Olson, L. B. Brubaker, S. R. Carpenter, M. M. Holland, S. P. Hubbell, S. A. Levin, J. A. MacMahon, P. A. Matson, J. M. Melillo, H. A. Mooney, C. H. Peterson, H. R. Pulliam, L. A. Real, P. J. Regal, and P. G. Risser. 1991. The sustainable biosphere initiative: An ecological research agenda. A report from the Ecological Society of America. *Ecology* 72: 371-412.

Lucas, O. W. R. 1991. *The Design of Forest Landscapes*. Oxford University Press, Oxford.

Lynch, K. M. 1996. A New Landscape Ecology Mapping Scheme for Coastal Environments: Galveston Island, Texas. M.S. Thesis. Texas A&M University, College Station.

MacDougall, A. and R. Turkington. 2005. Are invasive species the drivers or passengers of change in degraded ecosystems? *Ecology* 86: 42-55.

MacLean, A. S. and B. McKibben. 1993. *Look at the Land: Aerial Reflections on America.* Rizzoli International Publications, New York.

Mansourian, S., D. Vallauri, and N. Dudley. 2005. *Forest Restoration in Landscapes: Beyond Planting Trees*. Springer, New York.

Marsh, W. M. 1983. *Landscape Planning: Environmental Applications*. Addison-Wesley, Reading, MA.

Maslow, A. H. 1954. *Motivation and Personality*. Harper, New York.

Maslow, A. H. 2002. *The Psychology of Science: A Reconnaissance*. Harper & Row, New York.

Mattson, W. J. and N. D. Addy. 1975. Phytophagous insects as regulators of forest primary production. *Science* 190: 515-522.

McGarigal, K. and B. Marks. 1995. FRAGSTATS: Spatial Pattern Analysis Program for Quantifying Landscape Structure. PNW-GTR-351. U.S. Dept. of Agriculture, Forest Service. Pacific Northwest Research Station, Portland, OR.

McKinney, M., R. Schoch, and L. Yonavjak. 2007. *Environmental Science: Systems and Solutions.* Jones & Bartlett Publishers, Boston.

Millenium Ecosystem Assesment. 2005. *Ecosystems and Human Well-Being: Synthesis.* Island Press, Washington, D.C.

Miller, J. H. 2003. Nonnative invasive plants of Southern forests: A field guide for identification and control. U. S. Department of Agriculture, Forest Service, Southern Research Station, Asheville, NC.

Milton, N. R. 2007. *Knowledge Acquisition in Practice: A Step-by-Step Guide.* Springer, New York.

Milton, N. R. 2008. *Knowledge Technologies.* Polimetrica, Milan, Italy.

Mitchell, S. C. 2005. How useful is the concept of habitat? A critique. *Oikos* 110: 634-638.

Mladenoff, D. J. 2004. LANDIS and forest landscape models. *Ecological Modelling* 180: 7-19.

Mladenoff, D. J. and W. L. Baker. 1999. *Spatial Modeling of Forest Landscape Change: Approaches and Applications.* Cambridge University Press, Cambridge.

Moss, M. R. 2005. Toward fostering recognition of landscape ecology. In. J. A. Wiens and M. R. Moss (eds.), *Issues and Perspectives in Landscape Ecology.* Cambridge University Press, Cambridge.

Naiman, R. J. 1988. Animal influences on ecosystem dynamics. *BioScience* 38: 750-752.

Nassauer, J. I. 1997. *Placing Nature: Culture and Landscape Ecology.* Island Press, Washington, D.C.

National Research Council (U.S.). Committee on Assessing and Valuing the Services of Aquatic and Related Terrestrial Ecosystems. 2005. *Valuing Ecosystem Services: Toward Better Environmental Decision-Making.* National Academies Press, Washington, D.C.

National Research Council (U.S.) Rediscovering Geography Committee. 1997. *Rediscovering Geography: New Relevance for Science and Society.* National Academies Press, Washington, D.C.

Naveh, Z. and A. S. Lieberman. 1984. *Landscape Ecology: Theory and Application.* Springer-Verlag, New York.

Naylor, L. 2005. The contributions of biogeomorphology to the emerging field of geobiology. *Palaeoecology* 219: 35-51.

Naylor, L., H. Viles, and N. Carter. 2002. Biogeomorphology revisited: Looking towards the future. *Geomorphology* 47: 3-14.

Odum, E. P. 1953. *Fundamentals of Ecology.* Saunders, Philadelphia.

Odum, E. P. 1983. *Basic Ecology.* Saunders, Philadelphia.

Paine, R. T. 1966. Food web complexity and species diversity. *American Naturalist* 100: 65-75.

Paine, R. T. 1969. *Pisaster-Tegula* interaction: prey patches, predator food preference, and intertidal community structure. *Ecology* 50: 950-961.

Perry, G. and N. Enright. 2006. Spatial modelling of vegetation change in dynamic landscapes: A review of methods and applications. *Progress in Physical Geography* 30: 47-72.

Perry, G. and J. Millington. 2008. Spatial modelling of succession-disturbance dynamics in forest ecosystems: Concepts and examples. *Perspectives in Plant Ecology, Evolution, and Systematics* 9: 191-210.

Peterson, G. 2002. Contagious disturbance, ecological memory, and the emergence of landscape pattern. *Ecosystems* 5: 329-338.

Phillips, J. D. 1999. *Earth Surface Systems: Complexity, Order, and Scale.* Blackwell Publishers, Malden, MA.

Pickett, S. T. A. and M. Cadenasso. 2002. The ecosystem as a multidimensional concept: Meaning, model, and metaphor. *Ecosystems* 5: 1-10.

Pickett, S. T. A. and P. S. White. 1985. *The Ecology of Natural Disturbance and Patch Dynamics.* Academic Press, Orlando, FL.

Pimentel, D. 1963. Introducing parasites and predators to control native pests. *Canadian Entomologist* 95: 785-792.

Pinto, M., W. Rubink, R. Coulson, J. Patton, and J. Johnston. 2004. Temporal pattern of africanization in a feral honeybee population from Texas inferred from mitochondrial DNA. *Evolution* 58: 1047-1055.

Pinto, M., W. Rubink, J. Patton, R. Coulson, and J. Johnston. 2005. Africanization in the United States: Replacement of feral European honeybees (*Apis mellifera* L.) by an African hybrid swarm. *Genetics* 170: 1653-1665.

Polanyi, M. 1966. *The Tacit Dimension.* Doubleday, Garden City, NY.

Poole, G. C. 2002. Fluvial landscape ecology: Addressing the uniqueness within the river discontinuum. *Freshwater Biology* 47: 641-660.

Power, M., D. Tilman, J. Estes, B. Menge, W. Bond, L. Mills, G. Daily, J. Castilla, J. Lubchenco, and R. Paine. 1996. Challenges in the quest for keystones. *BioScience* 46: 609-620.

Raffa, K., B. Aukema, B. Bentz, A. Carroll, J. Hicke, M. Turner, and W. Romme. 2008. Cross-scale drivers of natural disturbances prone to anthropogenic amplification: The dynamics of bark beetle eruptions. *BioScience* 58: 501-517.

Reiners, W. A. and K. L. Driese. 2004. *Transport Processes in Nature: Propagation of Ecological Influences through Environmental Space.* Cambridge University Press, Cambridge.

Reynolds, K., M. Jensen, J. Andreasen, and I. Goodman. 2000. Knowledge-based assessment of watershed condition. *Computers and Electronics in Agriculture* 27: 315-333.

Ricketts, T. 2001. The matrix matters: Effective isolation in fragmented landscapes. *American Naturalist* 158: 87-99.

Ries, L., R. Fletcher, J. Battin, and T. Sisk. 2004. Ecological responses to habitat edges: Mechanisms, models, and variability explained. *Annual Review of Ecology, Evolution, and Systematics* 35: 491-522.

Ritter, D. F., R. C. Kochel, and J. R. Miller. 2002. *Process Geomorphology*. McGraw-Hill, Boston.

Roshier, D. and J. Reid. 2003. On animal distributions in dynamic landscapes. *Ecography* 26: 539-544.

Rudolph, D. C. and R. N. Conner. 1995. The impact of southern pine beetle induced mortality on Red-cockaded woodpecker cavity trees. In D. L. Kulhavy, R. G. Hooper, and R. Costa (eds.), *Red-Cockaded Woodpecker: Recovery, Ecology, and Management*. Stephen F. Austin State University, College of Forestry, Nacogdoches, TX.

Rykiel, E. J., R. N. Coulson, P. J. H. Sharpe, T. F. H. Allen, and R. O. Flamm. 1988. Disturbance propagation by bark beetles as an episodic landscape phenomenon. *Landscape Ecology* 1: 129-139.

Sakai, A., F. Allendorf, J. Holt, D. Lodge, J. Molofsky, K. With, S. Baughman, R. Cabin, J. Cohen, N. Ellstrand, D. McCauley, P. O'Neil, I. Parker, J. Thompson, and S. Weller. 2001. The population biology of invasive species. *Annual Review of Ecology and Systematics* 32: 305-332.

Sanderson, J. and L. D. Harris. 2000. *Landscape Ecology: A Top-Down Approach*. Lewis Publishers, Boca Raton, FL.

Scheller, R. and D. Mladenoff. 2007. An ecological classification of forest landscape simulation models: Tools and strategies for understanding broad-scale forested ecosystems. *Landscape Ecology* 22: 491-505.

Schneider, D. C. 1994. *Quantitative Ecology: Spatial and Temporal Scaling*. Academic Press, San Diego.

Shwiff, S., K. Gebhardt, K. Kirkpatrick, and S. Shwiff. 2010. Potential economic damage from introduction of brown tree snakes, *Boiga irregularis* (Reptilia: Colubridae), to the islands of Hawaii. *Pacific Science* 64: 1-10.

Solon, J. 2005. Incorporating geographical (biophysical principles) in studies of landscape systems. In J. A. Wiens and M. R. Moss (eds.), *Issues and Perspectives in Landscape Ecology*. Cambridge University Press, Cambridge.

Stallins, J. 2006. Geomorphology and ecology: Unifying themes for complex systems in biogeomorphology. *Geomorphology* 77: 207-216.

Stöglehner, G. 2003. Ecological footprint - a tool for assessing sustainable energy supplies. *Journal of Cleaner Production* 11: 267-277.

Stohlgren, T. and J. Schnase. 2006. Risk analysis for biological hazards: What we need to know about invasive species. *Risk Analysis* 26: 163-173.

Summerfield, M. 1991. *Global Geomorphology: An Introduction to the Study of Landforms*. Longman Scientific & Technical. Wiley, Harlow, UK.

Swank, W. T., J. B. Waide, D. A. Crossley Jr., and R. L. Todd. 1981. Insect defoliation enhances nitrate export from forest ecosystems. *Oecologia* 51: 297-299.

Swanson, F. J., T. K. Kratz, N. Caine, and R. G. Woodmansee. 1988. Landform effects on ecosystem pattern and processes. *BioScience* 38: 92-98.

Tansley, A. 1935. The use and abuse of vegetational concepts and terms. *Ecology* 16: 284-307.

Terrell, J., J. Hart, S. Barut, N. Cellinese, A. Curet, T. Denham, C. Kusimba, K. Latinis, R. Oka, J. Palka, M. Pohl, K. Pope, P. Williams, H. Haines, and J. Staller. 2003. Domesticated landscapes: The subsistence ecology of plant and animal domestication. *Journal of Archaeological Methods and Theory* 10: 323-368.

Tews, J., U. Brose, V. Grimm, K. Tielborger, M. Wichmann, M. Schwager, and F. Jeltsch. 2004. Animal species diversity driven by habitat heterogeneity/diversity: The importance of keystone structures. *Journal of Biogeography* 31: 79-92.

Thomas, M. 2001. Landscape sensitivity in time and space - an introduction. *Catena* 42: 83-98.

Tobler, W. R. 1970. A computer movie simulating urban growth in Detroit region. *Economic Geography* 46: 234-240.

Troll, C. 1939. Luftbildplan und ökologische bodenforschung. *Zeitschrift der gesellschaft für erdkunde zu Berlin* 7: 241-298.

Tufte, E. R. 1983. *The Visual Display of Quantitative Information*. Graphics Press, Cheshire, CT.

Tufte, E. R. 1990. *Envisioning Information*. Graphics Press, Cheshire, CT.

Tufte, E. R. 1997. *Visual Explanations: Images and Quantities, Evidence and Narrative*. Graphics Press, Cheshire, CT.

Tufte, E. R. 2006. *Beautiful Evidence*. Graphics Press, Cheshire, CT.

Turner, M. 2005a. Landscape ecology in North America: Past, present, and future. *Ecology* 86: 1967-1974.

Turner, M. 2005b. Landscape ecology: What is the state of the science? *Annual Review of Ecology Evolution and Systematics* 36: 319-344.

Turner, M. G., R. H. Gardner, and R. V. O'Neill. 2001. *Landscape Ecology in Theory and Practice: Pattern and Process*. Springer, New York.

Urban, D. L. 1993. Landscape ecology and ecosystem management. In W. W. Covington and L. F. DeBano (eds.), *Sustainable Ecological Systems: Implementing an Ecological Approach to Land Management.* USDA Forest Service GTM RM-247. U.S. Dept. of Agriculture, Rocky Mountain Forest and Range Experiment Station, Fort Collins, CO.

van der Valk, A. and B. Warner. 2009. The development of patterned mosaic landscapes: An overview. *Plant Ecology* 200: 1-7.

Vanslembrouck, I. and G. van Huylenbroeck. 2005. *Landscape Amenities: Economic Assessment of Agricultural Landscapes.* Springer, New York.

Veldkamp, A. and E. Lambin. 2001. Predicting land-use change. *Agriculture Ecosystems & Environment* 85: 1-6.

Viles, H. A. 1988. *Biogeomorphology.* B. Blackwell, Oxford, UK.

Wagner, H. and M-J. Fortin. 2005. Spatial analysis of landscapes: Concepts and statistics. *Ecology* 86: 1975-1987.

Werner, B. and D. McNamara. 2007. Dynamics of coupled human-landscape systems. *Geomorphology* 91: 393-407.

Wiens, J. A. 1989. Spatial scaling in ecology. *Functional Ecology* 3: 385-397.

Wiens, J. A. 1995. Landscape mosaics and ecological theory. In L. Hansson, L. Fahrig, and G. Merriam (eds.), *Mosaic Landscapes and Ecological Processes.* Chapman and Hall, London.

Wiens, J. A., C. S. Crawford, and J. R. Gosz. 1985. Boundary dynamics: A conceptual framework for studying landscape ecosystems. *Oikos* 45: 421-427.

Wiens, J. A. and M. R. Moss. 2005. *Issues and Perspectives in Landscape Ecology.* Cambridge University Press, Cambridge.

With, K. 2002. The landscape ecology of invasive spread. *Conservation Biology* 16: 1192-1203.

Wright, J. and C. Jones. 2006. The concept of organisms as ecosystem engineers ten years on: Progress, limitations, and challenges. *BioScience* 56: 203-209.

Wu, J. and J. David. 2002. A spatially explicit hierarchical approach to modeling complex ecological systems: Theory and applications. *Ecological Modelling* 153: 7-26.

Wu, J. and R. J. Hobbs. 2007. *Key Topics in Landscape Ecology.* Cambridge University Press, Cambridge.

Xi, W., R. N. Coulson, A. G. Birt, S. Zongbo, J. D. Waldron, C. W. Lafon, D. M. Cairns, M. D. Tchakerian, and K. D. Klepzig. 2009a. Review of forest landscape models: Types, methods, development and applications. *Acta Ecologica Sinica*: 69-78.

Xi, W., J. D. Waldron, C.W. Lafon, D. M. Cairns, A.G. Birt, M. D. Tchakerian, R. N. Coulson, and K.D. Klepzig. 2009b. Modeling long-term effects of altered fire regimes following southern pine beetle outbreaks (North Carolina). *Ecological Restoration* 27: 24-26.

Zonneveld, I. S. 1995. *Land Ecology: An Introduction to Landscape Ecology as a Base for Land Evaluation, Land Management and Conservation.* SPB Academic Publishing, Amsterdam.

Zonneveld, I. S. and R. T. T. Forman. 1990. *Changing Landscapes: An Ecological Perspective.* Springer-Verlag, New York.

Index

A

Abies religiosa. See oyamel fir.
abiotic storage 52, 156.
adaptive cycle 53, 55, 56, 57, 59.
Adelges tsugae. See hemlock woolly adelgid.
aeolian processes 173.
African honey bee 19, 20, 21.
age structure 18.
agriculture 63, 109, 217, 220, 233.
Agrilus panipennis. See emerald ash borer.
allogenic engineers 193.
Alsophilia pometaria. See fall cankerworm.
A. m. scutellata. See African honey bee.
analysis 92, 107, 239, 240, 243.
angular geometric pattern 234, 235.
angular geometry 234, 235.
anthropogenic ecotones 146.
Apis mellifera 19, 38.
application 9.
appraisal 255, 257.
architecture 86, 87.
Arthropoda 52.
Asclepiadaceae. *See* milkweed.
ash tree 78, 206.
attrition 210, 221, 223.

autocorrelation 242, 245.
autogenic engineers 193, 194.
autotrophic 50.

B

beavers 75, 77, 101, 193, 195.
balsam woolly adelgid 74.
behavior 70, 97, 99, 150, 152, 153.
bioconstruction 189, 191.
biodiversity 211.
bio-ecology tradition 6, 9.
bioerosion 189, 190.
biogeomorphology 75, 185, 187, 189.
biomass 43, 245.
bioprotection 189, 190.
biosphere 12.
Boiga irregularis. See brown tree snake.
boundary
   hard boundaries 228.
   soft boundaries 228.
Brassicaceae. *See* mustard.
Brown tree snake 200, 201.

C

capital 55, 58.
carnivore 50, 99.
Carolina hemlock 206.
carrier 73, 162, 175.

cartography 40.
Castor canadensis. See beavers.
categorical maps 240, 243.
chorology (choric) 148, 149.
civil engineering 87.
coastal processes 174, 179.
colluvial transport 174.
commerce 136, 233.
commodity 218.
community 12, 18, 28.
composition 123, 125, 221, 226, 243.
condition 66, 75, 97, 98, 99.
configuration 91, 104, 243, 244.
connectedness 55, 229.
connectivity 26, 123, 141, 142, 144, 229, 230, 242.
consequence 16, 73, 145, 163, 164, 177, 183, 208.
consumption 50, 53, 63.
corridor 67, 69, 121.
  corridor attributes 122.
    composition 123.
    length 123.
    width 122.
  corridor function 122, 133.
    conduits 134.
    sinks 137, 140.
    sources 137.
  corridor origin 122, 129.
    disturbance corridor 129.
    environmental corridor 129.
    introduced corridor 129.
    regenerated corridor 129.
    remnant corridor 129.
  corridor types 122.
    line corridors 125.
    stream corridors 125.
    strip corridors 125.
    tube corridors 127.
Coweeta Hydrological Laboratory 30, 31, 32.

D

Danaus plexippus. See monarch butterfly.

data 168, 169, 170, 239, 245.
decomposition 52, 53, 63.
definition 6, 109, 255.
Dendroctonus frontalis. See southern pine beetle.
detritivores 52, 53, 99.
development 7, 9, 189, 230.
digital age 7.
disease epidemiology 177.
dispersal 14, 152, 153, 204.
dissection 119, 221, 222, 223.
disturbance 29, 75, 77, 113, 208.
  disturbance regime 212.
  natural disturbance 24, 75, 76, 77.
domains of synthesis 85.
driver 186, 216, 217, 219, 220.

E

earthworms 195, 196.
Eastern hemlock 206.
ecological footprint 234.
ecological thoughts 104.
ecology 5, 11, 19, 43, 65.
ecoregion 12, 13, 16, 47, 86.
ecosystem 12, 47, 48, 49, 53, 59.
  ecosystem cluster 96, 147, 148.
  ecosystem engineering 113, 191, 192, 204.
  ecosystem engineers 75, 77, 189, 192.
  ecosystem function 49, 56, 60.
  ecosystem health 63.
  ecosystem management 48, 61, 63.
  ecosystem services 48, 59, 60, 61, 62.
  meta-ecosystems 148.
Ecosystem Management Decision Support System (EMDS) 258.
ecotone 69, 72, 144, 145, 146.
ecotope 3, 49.
edge 69, 144, 227.
  edge species 118, 228.
  high contrast 228.
  low contrast 228.
edge effects (edge influences) 227.

Elaphe spp. *See* rat snakes.
elicitation 256.
emerald ash borer 77, 78, 206.
emergent properties 11, 16, 18.
endangered species 23, 211, 250.
energy 99, 168, 169.
energy flow 18, 52, 53.
energy subsidies 234.
entities 73, 160, 162, 167, 174.
environment 95, 97, 149, 167, 172, 177.
environmental domain 72, 164.
environmental space 162, 164.
ephemeral 116, 117, 118.
epicenters 25.
episode 182, 183.
epizootic 29, 31, 33.
experiential knowledge 255, 257.
explicit knowledge 256.
extent (range) 36, 37.

F

fall cankerworm 30, 31, 33.
fitness 103.
fluvial processes 174.
forestry 2, 63, 210, 233.
fragmentation 24, 52, 210, 221, 222, 223, 227.
FRAGSTATS 21, 243.
Fraxinus spp. *See* ash tree
functional connectivity 151, 229, 242.
functional heterogeneity 26, 27, 28, 70, 96, 149.

G

geocomplex 148.
geo-ecology tradition 6.
geographic range 154.
geometry 4, 66, 95, 104.
geomorphology 7, 86, 172.
geostatistics 245.
global climate change 187, 233.
Glossina spp. *See* tsetse fly.

gradients 188, 245.
grazing 220, 233.
grizzly bear 212.

H

habitat 24, 25, 26, 100, 101, 102.
  habitat connectivity 144, 222, 229.
  habitat heterogeneity 145, 226, 227, 229.
  habitat places 226.
habitat targets 25, 101.
health 61, 62, 199.
hemlock woolly adelgid 54, 206.
herbivore 28, 31, 50, 99, 100, 197.
heterogeneity 3, 18, 84, 232, 241.
heterotrophs 50.
heuristic knowledge 240, 255, 257.
hierarchy 11, 16, 18.
home range 25, 135, 152, 154.
Homo sapiens 150.
honey bees 19, 20, 21, 22, 41, 99.
human settlement 216, 234, 237.

I

IALE vii.
impact 29, 30, 71, 209.
  ecological impact 210.
  economic impact 209.
  political impact 209.
  social impact 209.
individual 12, 13, 18.
information 168, 169, 170.
initiating cause 73, 161, 162, 163.
integrity 63.
interior species 118, 228.
internal entity 112, 123, 124.
invasive species 77, 189, 199, 200, 201.

K

k-strategist 118
keystone species 20, 75, 189, 195.
knowledge 169.
  knowledge acquisition 255.

knowledge engineering 254.
knowledge integration 257.
knowledge-based system 258.
kudzu 202.

L

land-cover 146, 187.
landform 75, 86, 172, 187.
landscape 2, 3, 5, 12, 81, 85, 86.
  landscape change 73, 185, 221.
  landscape-cover change 65, 75, 185, 186, 187.
  landscape domestication 77, 116, 216.
  landscape ecology 6.
  landscape environment 95, 97, 149, 167.
  landscape function 71, 159.
  landscape management 66, 79.
  landscape mosaic 86, 87, 148, 149.
  landscape pattern indices (LPIs) 240, 244.
  landscape scale 156.
  landscape structure 66, 95, 104.
  landscape-use change 74, 77, 216.
landschap 81.
land-use 146, 187.
large-scale 41.
levels of integration 11, 12, 13, 14.
life cycle 150.
life history 150.
lightning 25, 58, 163, 207, 214.
loblolly pine 22, 25.
longleaf pine 22, 25, 26, 28.
lovebugs 137, 140.
LULC 187.
Lumbriculidae. *See* earthworms.

M

maintenance movement 152, 154.
mangroves 191.
map 8, 37, 40.
map symbology 147.
Mariposa Monarca Biosphere Reserve 120.
mass movement 174, 178.
mass wasting 174.
materials 100, 170, 171.
matrix 69, 70, 141, 143, 230.
  high-resistance matrix 142.
  low-resistance matrix 142.
  matrix permeability 142.
Meleagris gallopavo. See wild turkey.
mesh size 136.
metrology 43.
migration 119, 120, 153.
migration corridors 135.
milkweed 120.
minimal mapping unit 243.
monarch butterfly 119, 120, 121.
mosaic 3, 148, 230.
movement 152.
mustard 113, 114.
Myocastor coypus. See nutria.

N

n-dimensional hypervolume 103.
neighborhood mosaics 148.
neighborhoods 243.
network form 136.
niche 101, 102.
  fundamental niche 103.
  niche breadth 102.
  niche overlap 103.
  realized niche 103.
nutria 201, 202, 203.
nutrient cycling 18, 52, 171, 211.
  environmental phase 171.
  organismal phase 171.

O

object oriented design (OOD) 250.
Odocoileus virginianus. See white-tailed deer.
Opuntia spp. *See* prickly pear cactus.
Ostreidae. *See* oysters.
overprinting 234.
oyamel fir 120.

oysters 191.

P

palimpsest 234, 237.
panarchy 57, 58, 59, 212.
patch 67, 107.
  patch attributes 107.
    patch location 112.
    patch number 109.
    patch shape 109.
    patch size 108.
  patch function 117.
  patch origin 113.
    disturbance patch 113.
    environmental patch 116, 117.
    ephemeral environmental patch 118.
    introduced patch 116.
    regenerated patch 116.
    remnant patch 113.
pattern 3, 148, 230.
pattern/process paradigm 3, 159.
perception 66, 95, 149.
perforation 221, 222, 223, 227.
perturbation 208.
philopatry 153.
physiography 86, 172.
Picoides borealis. See Red-cockaded woodpecker.
pine 20, 22, 25, 29, 116, 129, 181, 197, 251, 252.
Pinus spp. *See* pine.
Pinus echinata. See shortleaf pine.
Pinus elliotti. See slash pine.
Pinus palustris. See longleaf pine.
Pinus taeda. See loblolly pine.
pixel 37, 38, 39.
place 85, 86.
Plecia nearctica. See lovebugs.
population 12, 13, 14.
post oak savanna 50, 67, 104, 105.
potential 58, 59.
Prairie pothole. See wetland slough.
prickly pear cactus 202.

primary production 49, 50, 53.
primary productivity 50, 210.
process 53, 159.
process modeling 249.
productivity 43, 55.
project 255, 258.
propagation vector 73, 161, 162, 164.
Pueraria montana. See Kudzu.

Q

qualitative information 255.
quantitative data 92, 255.

R

r-strategist 118
radiant energy 99.
rare species 211.
rat snakes 25.
red-cockaded woodpecker 20, 22, 23, 24, 212, 243.
red imported fire ant 106.
relative abundance 14, 18.
remember 58, 59.
representative fraction (RF) 40, 42.
resilience 56, 58, 63.
resin volatiles 25.
resolution (grain) 37, 39, 243.
resource 97, 99, 100.
revolt 58, 59.
Rhizophora spp. *See* mangroves.
road density 136.
road ecology 135, 136, 140.

S

scalar 168.
scale 4, 35, 40, 43.
  spatial 36, 243.
  temporal 36, 243.
scenarios 181, 252, 253.
shape 81, 108, 243, 244.
shifting mosaic 203, 233.
shortleaf pine 22.
shrinkage 210, 222, 223.
silvicultural practices 252.

sink 108, 119, 137, 140.
site 49, 86.
SI units 43.
  derived 43.
slash pine 22.
sleeping sickness 197.
small-scale 41.
Solinopsis invicta. See red imported fire ant.
source 108, 119, 122, 137, 140.
southern pine beetle 22, 23, 24, 25, 129, 131, 197, 243, 250, 252, 253.
southern yellow pines 22, 197.
space 7, 36, 37, 43, 85.
spatially explicit 3, 10, 241, 248.
spatially explicit models 248, 253.
spatial statistics 240, 241, 245.
species-area relationship 226.
species diversity 18, 211.
species richness 18.
stepping stone 25, 108, 119, 144.
Streamside Management Zones (SMZs) 137, 139.
stress 208.
stressor 208.
sub-system 16, 18.
supra-system 16, 18.
sustainability 63.
symbology 146, 147.
synthesis 239, 248.
system 16, 17, 18.

T

tacit knowledge 256.
TAMBEETLE 251, 252, 253.
T. caroliniana. See Carolina hemlock.
thermodynamics
  first law of thermodynamics 52.
  second law of thermodynamics 52.
topology (topic) 148.
transformation 210, 211, 223, 224, 236.
transport model 160, 163, 164, 177, 182, 183.

Transvolcanic Plateau 120.
Trypanosoma spp. 197.
tsetse fly 197, 198.
Tsuga canadensis. See Eastern hemlock.
Tsuga caroliniana. See Carolina hemlock.

U

urbanization 146, 210, 233.
urban planning 87.
Ursus arctos horribilis. See grizzly bear.
US-IALE 1.

V

vector 73, 163, 172.
visualization 251, 252.

W

wetland slough 113, 114.
white-tailed deer 101, 104, 105.
wild turkey 212.

Z

zoogeomorphologists 66, 75, 189.